D1046916

Conductive Polymers and Plastics

Conductive Polymers and Plastics

Edited by James M. Margolis

Chapman and Hall

NEW YORK AND LONDON

First published 1989
by Chapman and Hall
29 West 35th Street, New York, NY 10001

Published in Great Britain by
Chapman and Hall, Ltd.
11 New Fetter Lane, London EC4P 4EE

Printed in The United States Of America

Library of Congress Cataloging in Publication Data

Conductive polymers and plastics handbook.

Includes index.
1. Plastics—Electric properties. 2. Organic conductors/ I. Margolis, James M.
TP1122.C66 1989 620.1'92397 88-25816
ISBN 0-412-01431-9

British Library Cataloguing in Publication Data

Conductive polymers and plastics handbook.
1. Conductive polymers
I. Margolis, James M.
620.1'9204297

ISBN 0-412-01431-9

Dedicated to my wife, Rena

Contents

Part I

Polymers

Chapter 1

Electrically Conductive Polymers

John R. Reynolds, Charles K. Baker, Cynthia A. Jolly, Paul A. Poropatic, and Jose P. Ruiz

Introduction

The 1977 report that polyacetylene,[1] a conjugated organic polymer, could attain high levels of electronic conductivity when oxidized by suitable reagents initiated a significant research thrust that has included the participation of chemists, physicists, and material scientists. Over the last decade, literally thousands of technical papers have been published in this field, which has now been expanded to include a multitude of polymer systems. In fact, a book entitled *Polyacetylene: Chemistry, Physics and Materials Science* has been published detailing significant research effort that has been dedicated to this polymer alone, and a set of handbooks treats the field in more detail.[2]

The concept of conductivity and electroactivity of conjugated polymers was quickly broadened from polyacetylene to include a number of conjugated hydrocarbon and aromatic heterocyclic polymers, such as poly(*p*-phenylene),[3] poly(*p*-phenylene vinylene),[4] poly(*p*-phenylene sulfide),[5] polypyrrole,[6] and polythiophene.[7] An all-encompassing list would be quite extensive.

In this chapter we address the synthesis and electronic conductivity properties of conjugated organic polymers, using specific examples to illustrate our points. Further, we describe how transition metal ions, with suitably conjugated ligands, can be used to introduce charge carriers into polymers. Lastly, we describe some of the current and potential applications of these materials as they now approach a marketable form.

Polyacetylene: The Prototype Conducting Polymer

Polyacetylene, $(CH)_x$, a simple conjugated polymer, can be synthesized by a variety of routes. Most research on $(CH)_x$ as a conductive polymer has been performed on what is known as the "Shirakawa" type,[1] which is synthesized via the Ziegler–Natta polymerization of acetylene shown in Equation 1. The polymer forms as a highly crystalline, completely insoluble, mat of fibrils.

$$\text{H-C}\equiv\text{C-H} \xrightarrow[\text{AlEt}_3]{\text{Ti(OBu)}_4} \mathbf{1} \xrightarrow{\Delta} \mathbf{2} \quad (1)$$

These fibrils have diameters on the order of 50–200 Å, which facilitates redox reactions used to impart conductivity to this polymer. At low temperature the polymer is synthesized as the *cis*-transoid isomer (1), which is easily thermally converted to the more stable *trans*-transoid form (2) as shown.

To obtain materials of a more controllable morphology, soluble precursor systems have been developed that, on elimination of some small molecule, leave the fully conjugated $(CH)_x$ chain.[8,9] This is illustrated in Equation 2 for a polymerization route developed by Feast. The elimina-

$$\xrightarrow[\text{Me}_4\text{Sn}]{\text{WCl}_6} \quad \xrightarrow{\Delta} \quad + \quad (2)$$

tion is carried out thermally and thus the polymer is obtained in the *trans*-transoid form by this method.

As synthesized, *trans*-$(CH)_x$ is a semiconductor having an electrical conductivity on the order of 10^{-5}–10^{-6} ohm^{-1} cm^{-1}. The actual bond lengths alternate, half having more single-bond character and half having more double-bond character, as opposed to equal length if the system were perfectly conjugated. The polymer's structure is stabilized, via a phenomenon similar to a Peirels transition, leading to bond alternation.

The introduction of charged carriers is accomplished by redox reactions, which are commonly denoted as "doping reactions" in that they increase the conductivity of the polymer in a manner analogous to the doping of inorganic semiconductors. Utilizing oxidative doping as an example, removal of an electron from the polymeric π system, as shown in Equation 3, leads to a delocalized radical ion. This radical ion is viewed as a polaron (3) in that it represents a charged and paramagnetic defect in the $(CH)_x$ lattice. Also contained within the synthesized $(CH)_x$ are a number of neutral paramagnetic defects termed solitons (4). Oxidation

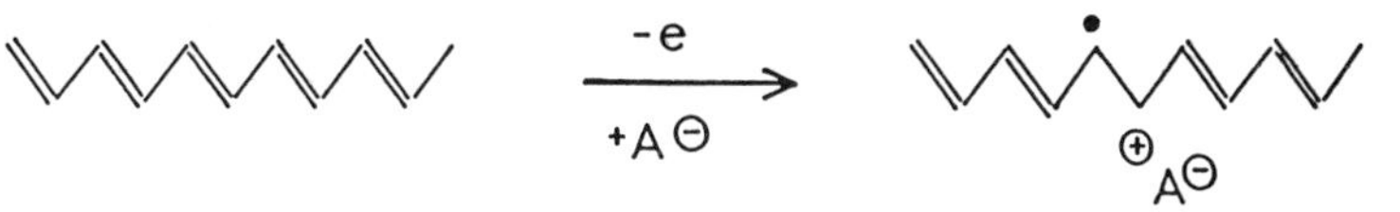

-e
+A⊖
⊕A⊖
3
(3)

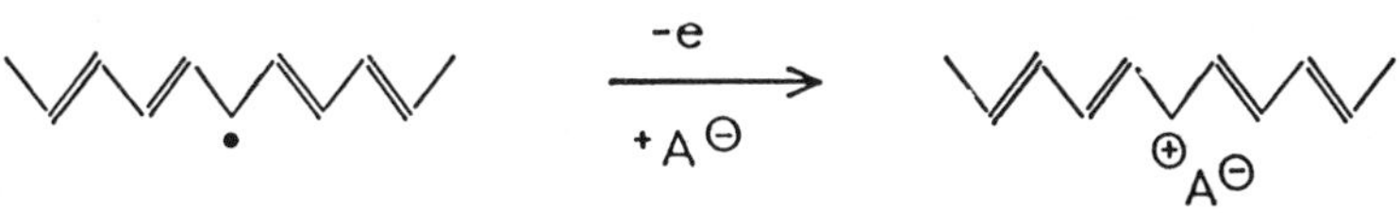

-e
+A⊖
⊕A⊖
4
5
(4)

of a neutral soliton yields a charged soliton (5), which is also delocalized along the polymer chain. A second oxidation of a chain containing a polaron, followed by radical recombination, yields two charged carriers on each chain (6), as illustrated in Equation 5.

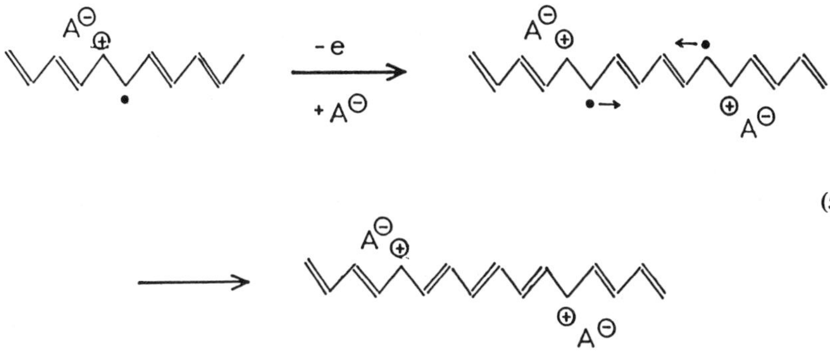

A⊖⊕
-e
+A⊖
A⊖⊕
⊕A⊖
(5)
A⊖⊕
⊕A⊖
6

These charged sites on the polymer backbone must be charge compensated, and during the doping process anions penetrate into the polymer matrix for this purpose. If the doping is carried out chemically, then the dopant anions are generated from the oxidizing agent, and if the doping is carried out electrochemically, an electrolyte anion is incorporated as the dopant anion. Chemical and electrochemical reduction can also be used in a manner analogous to the preceding to yield polyacetylene with negatively charged (*n*-type) carriers.

The delocalized charges on the polymer chain are mobile, not the dopant ions, and are thus the current-carrying species for conduction. These charges must hop from chain to chain, as well as move along the chain, for bulk conductivity to be possible. The actual mechanisms of conduction are quite complicated and still under debate, as will be discussed later. The preceding, relatively simplistic picture of charge creation in $(CH)_x$ is given because it can serve as a model for understanding conductivity in the systems to be discussed further.

Two of the major drawbacks to the use of $(CH)_x$ in practical applications are extreme air sensitivity and lack of processability. In the neutral, semiconductive form it irreversibly oxidizes, under exposure to air, degrading the π system along the backbone. The highly conducting doped polymer is also reactive in air, probably via a series of reactions involving moisture, with a corresponding precipitous conductivity drop. $(CH)_x$ is also completely insoluble and infusible, which limits its ability to be used in typical polymer film and fiber operations. For this reason the bulk of this chapter is directed to describing conductive polymer systems in which these limitations are not so severe.

Electrochemical Polymerization of Aromatic Molecules

Electrochemical oxidation of resonance-stabilized aromatic molecules has become one of the principal methods used to prepare conjugated, electronically conducting polymers. Since the first reports that the oxidation of pyrrole at a platinum electrode in the presence of a supporting electrolyte will produce a free-standing electrically conductive film,[6,10,11] many other aromatic systems have been similarly electropolymerized. These include thiophene,[7,12–14] furan,[7] carbazole,[15,16] azulene,[7,17] indole,[7] aniline,[18–21] selenophene,[22–24] phenol,[25,26] and thiophenol,[27] as well as many substituted, multiring, and polynuclear aromatic hydrocarbon systems such as 3-methylthiophene,[28,29] bithiophene,[30] and pyrene.[31,32] All the resulting polymers have a conjugated backbone and are oxidized electrochemically as they are formed on the electrode from a solution containing both electrolyte and monomer. This oxidation leads to the incorporation of charge-compensating anions into the oxidized film. These anions are thus the dopant ions. As in $(CH)_x$, the resulting polymer films are electrically conducting when oxidized and electrically insulating when neutral.

The two most studied members in this group of polymers are polypyrrole (7)[33–37] and polythiophene (8).[38,39] For illustrative purposes we will

7 8

detail the data concerning their electropolymerization and electrochemical characterization in this section. It can be expected that other, comparable monomer systems will exhibit similar behavior. Figure 1.1 shows a typical cyclic voltammogram of pyrrole. Several important features should be noted. On the initial anodic sweep there is a single peak at 0.9 V versus the standard calomel reference electrode, which corresponds to the oxidation of pyrrole at the electrode surface. The lack of a reductive peak on the return sweep indicates that the oxidation is irreversible, and the highly unstable radical-cations formed during oxidation undergo an immediate chemical reaction. If additional cyclic voltammograms are performed on the same untreated electrode in the same electrolyte/monomer solution, the monomer oxidation peak current will continue to grow

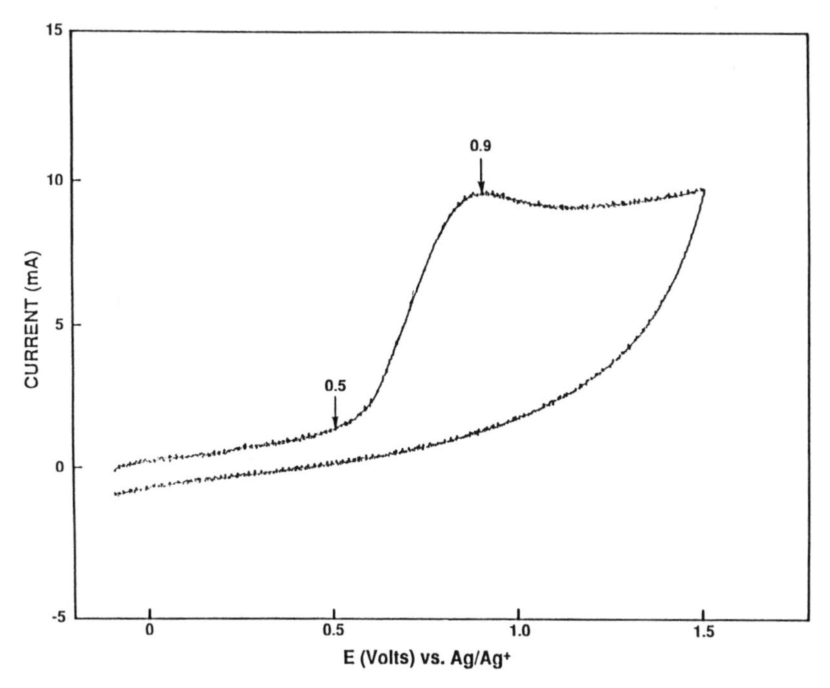

Figure 1.1 Cyclic voltammogram of pyrrole; scan rate = 25 mV s^{-1}

to progressively larger values, indicating that the polypyrrole produced remains on the electrode surface as it is produced. This also implies that the polypyrrole is produced with a high enough conductivity to participate in subsequent monomer oxidation and polymerization. Once a sufficient amount of polymer is deposited on the electrode, peaks due to polymer redox processes appear in the cyclic voltammogram.

Several studies of the electropolymerization of pyrrole and thiophene[35–49] onto a number of supporting electrode substrates have been made in an attempt to elucidate both the mechanism during the initial stages of polymer film formation and the role that experimental conditions play in the mechanism. These studies have used a number of electrochemical techniques, including cyclic voltammetry, chronocoulometry, and chronoamperometry as well as optical absorption and gravimetric techniques. The cyclic voltammogram shown in Figure 1.2 is similar to the one shown in Figure 1.1 except that the switching potential is close to the monomer peak potential. In this experiment, a nucleation loop appears at the foot of the oxidation wave. This nucleation loop is characterized by a potential region in which the current on the reverse sweep is higher than that seen on the forward sweep and is evidence for the formation of polymer at the electrode surface. On subsequent scans, the foot of the oxidation wave starts at slightly lower potentials, suggesting that polymerization occurs more readily on polypyrrole than on the underlying metallic substrate.

Another electrochemical method used to study the nucleation and

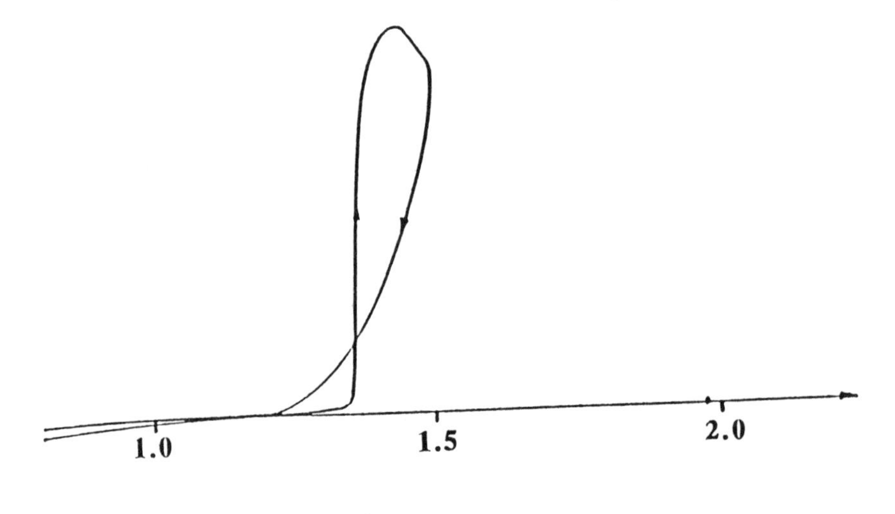

Figure 1.2 Cyclic voltammogram of pyrrole showing nucleation loop.

growth mechanism is chronoamperometry. In this technique, the potential at the working electrode is stepped from a rest potential to a potential where oxidation can occur and current is measured as a function of time. The current–time transient shown in Figure 1.3 is typical for the oxidation of pyrrole or thiophene and contains three distinct regions of behavior indicative of three separate stages involved during film formation. The first stage is characterized by a sharp initial current spike whose decay is potential dependent, nonexponential, and of time scales longer than that expected if it were due solely to charging of the ionic double layer at the electrode surface. Rotating ring–disc electrode and coulometric experiments of thiophene deposition on gold[37] suggest that this initial current spike has contributions from both a nucleation and growth process as well as from the formation of soluble oligomers. Quantitative analysis of the amount of charge passed and the duration of this spike suggest that this first stage corresponds to the coverage of the metallic electrode by a one-monolayer-thick polymer film. A separate study[40] has found similar results regarding monolayer coverage by using Tafel slope analysis and measurement of Pt electrode surface coverage by adsorbed pyrrole.

The beginning of the second stage is marked by a minimum in the

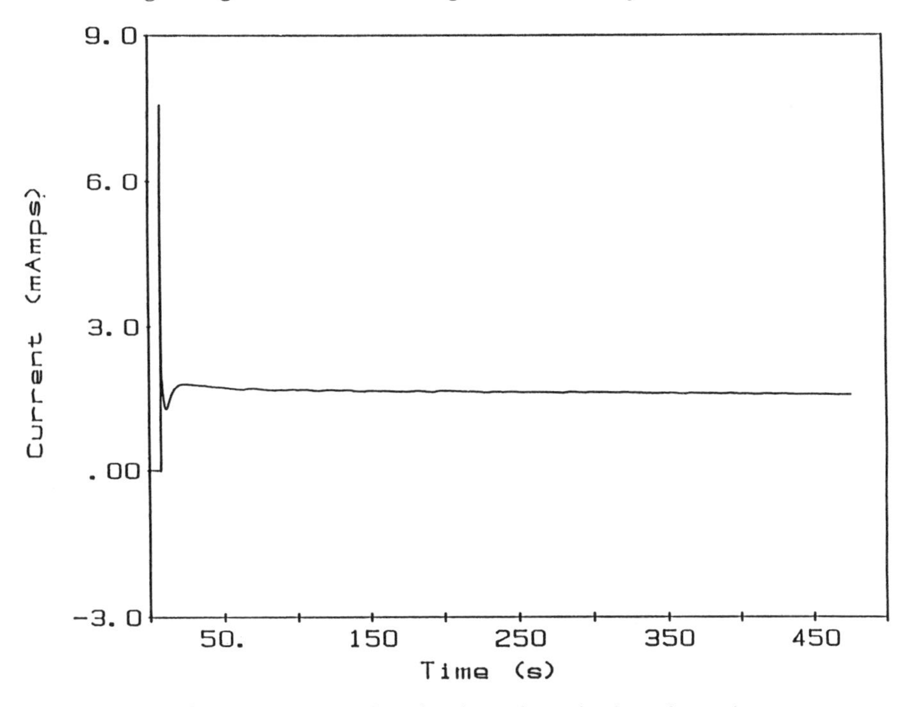

Figure 1.3 Current-time transient for the polymerization of pyrrole.

current–time transient followed by a rising current that levels off relatively quickly. During this stage, the rise in current is proportional to t^2, which indicates that the growth of the surface phase is determined by the rate of electron transfer at the surface of the expanding phase. This is characteristic of instantaneous formation of nucleation sites on the polymer monolayer followed by a three-dimensional growth until these sites overlap. The beginning of the third stage corresponds roughly with this overlap of nucleation sites and additional growth perpendicular to the underlying substrate. The current response in this region is independent of time, when diffusional constraints are absent, which is generally the case over the bulk of film formation. The fact that polymerization in this region is not diffusion limited is confirmed by the linear charge–time,[42,44,49] mass–time,[42,49] and absorbance–time[44] results obtained from chronocoulometric, chronogravimetric, and spectroelectrochemical measurements, respectively.

Other theoretical and electrochemical studies involving the relationship between electropolymerization conditions and the properties of the resulting film have led to conclusions about the efficiency and mechanism of the oxidation/deposition process. The picture that emerges from these studies for the polymerization during the bulk of film formation is shown in Scheme 1 for polypyrrole.

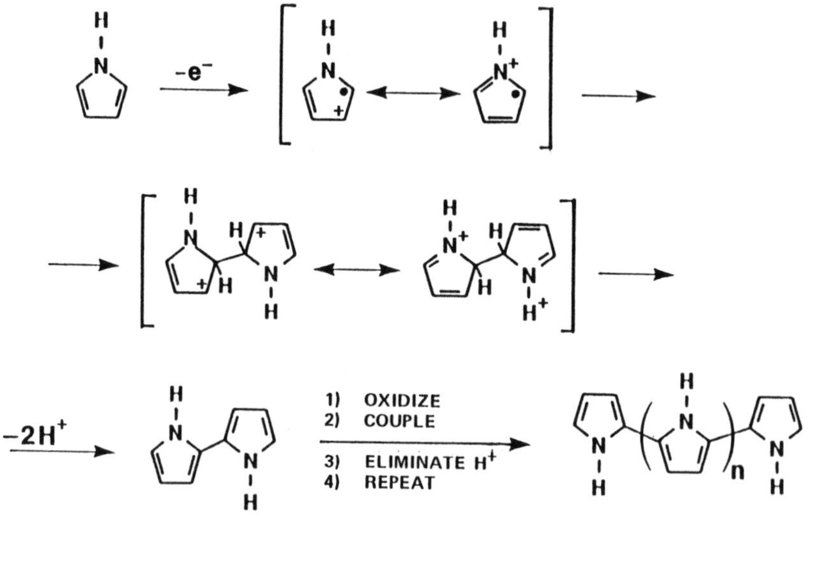

$-e^-$

$-2H^+$

1) OXIDIZE
2) COUPLE
3) ELIMINATE H^+
4) REPEAT

Scheme 1

Polymerization is initiated by the oxidation of monomer to yield a radical-cation species. Although some[44] have suggested that the next step may be formation of a covalent bond between this radical cation and a neutral monomer, followed by an electron transfer from this product, this seems unlikely because of the potential dependence found in the copolymerization of pyrrole with substituted pyrroles.[50] Therefore, the second step is more likely to be the coupling of radical cations to yield dimers. These resulting dimers can then be oxidized at a lower potential[51–53] to yield a radical-cation dimer in which the highest spin density is found at the terminal carbons alpha to the pyrrole nitrogens. These dimer radical ions (or at later stages, oligomers) may couple with another radical-ion monomer, dimer, or oligomer species, eventually precipitating and leading to an insoluble film. This lack of solubility is inherent to unsubstituted polymers containing a polyconjugated planer backbone. This planarity is due to the energy minima associated with the overlap of adjacent π-orbitals that is necessary to achieve efficient electron transport along the polymer chain. The insolubility of these materials limits the number of analytical techniques that can be employed for characterization.

In practice, the highest-quality electrically conducting polymer films are prepared in an inert atmosphere using a three-electrode electrochemical cell. The reaction mixture usually includes 0.01 to 0.2 molar pyrrole in an unreactive aprotic solvent and an electrolyte such as tetraethylammonium tosylate, tetrabutylammonium tetrafluoroborate, or lithium perchlorate. In addition, a small percentage of water, typically 2%, can be added, as it has been shown to enhance the mechanical properties of the resulting film.[54] The water is thought either to provide a labile counter electrode reaction or to participate in the initial stages of film formation, although the exact mechanism of this enhancement is still unclear.

Variation of the identity of the dopant or counter anion used in the synthesis provides a means to modify chemically the physical, electrochemical, and conductivity properties of these polymer films. A number of papers[29,55–59] have treated this topic. One study[46] has found that films with superior mechanical properties (smoothness and flexibility) possessed higher conductivities. These mechanical properties were found to be related to anion size, with the larger organic atoms (i.e., large amphiphilic aromatic sulfonates such as dodecylsulfonate) producing higher-quality films by increasing the order of the polymer chains during the polymerization and deposition process. This study also found a correlation between the acid strengths of the dopant anions and conductivity, with weak acid anions giving lower conductivities. The cyclic voltammograms of thin polypyrrole films prepared and characterized in different anion-containing solutions were used as an estimate of film quality by examining the sharpness of the reduction peak on the first scan. Again, the larger organic dopants gave higher-quality electrochemical responses in addition to higher conductivities.

Similar studies[58] on 3-methyl thiophene concluded that the nature of the anion plays a primary role in determining the structure of the polymer during synthesis. The nature of the cation affects the electrochemical behavior of the polymer but not its structure.

Other experimental variables affecting the properties of the electrochemically deposited conducting polymers include different methods of electrochemical synthesis—i.e. constant potential, constant current, and potential sweep, solvent choice, atmospheric conditions, e.g., N_2 or air, and electrode material (e.g., Pt, Au, indium-tin oxide, and glassy carbon).

Electrochemistry of Conducting Polymer Films

The electrochemical properties of organic conducting polymers are critical to any potential electrode applications. When the polymer films are left in contact with the substrate on which they were formed, they can be cycled repeatedly between the oxidized and neutral forms. In general, the oxidized form of the polymer is highly conducting because of the formation of polarons and bipolarons and insulating in the reduced, or neutral, state. Several common electrochemical techniques are used to examine the nature of the oxidation or doping process that controls the level of conductivity—namely, cyclic voltammetry and chronocoulometry—because of their ease of application and ability to control accurately the oxidation state of the polymer.

The cyclic voltammogram produced from a polypyrrole film in a monomer free supporting electrolyte shown in Figure 1.4 contains a peak for the oxidation of the film at ca. -0.1 V on the anodic sweep and a corresponding reduction peak at ca. -0.3 V on the cathodic sweep. The peak current values for the anodic process, i_{pa}, scale linearly with the sweep rate as expected for electron transfer to surface-bound species. The peaks are also displaced toward higher positive potentials with an increase in sweep rate. The peak currents for the reductive process are much broader and are difficult to analyze. This has been attributed to inhomogeneities of the polypyrrole in the potential regions of interest.[60] Although this redox reaction is electrochemically reversible over many scans, the peak shapes and separation indicate that this process deviates from a strictly Nernstian behavior and has kinetic limitations.

Several works appear in the literature that attribute the non-Nernstian behavior of the polymer redox process to ion mobility, as seen in the unsymmetrical appearance of the cyclic voltammogram. Early work[56,61]

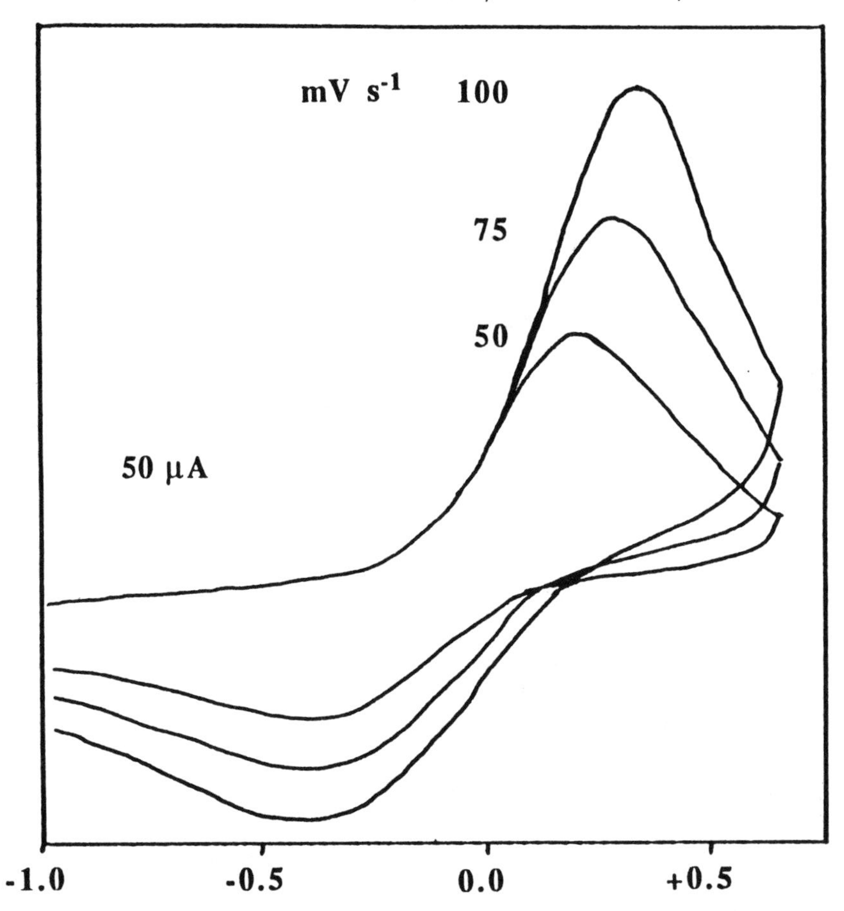

Figure 1.4 Cyclic voltammogram for polypyrrole at varied scan rates.

found significant changes in the shifts of the E_{pa} values and peak shapes as a function of anion identity that was attributed to changes in the kinetics of the redox process. Experiments in which only the anion was replaced during analysis of the polymer film suggested that any anion-dependent differences in peak characteristics occur primarily as a result of the electrochemical synthesis.[58] They also found that the electrochemical characteristics of the film are enhanced as the size of the anion used during synthesis increases. Further investigations brought to light the effects of the cation in the polymer redox process.[58] By replacing the cation in the electrolyte used for postsynthesis characterization of the polymer, they found that changes in the cation affect the electrochemical doping–undoping processes of the polymer. Gravimetric studies of polypyrrole using the sensitive quartz crystal microbalance indicate that doping is accompanied by either anion insertion (as intuitively expected) or cation exclusion from the polymer matrix, depending on the electrolyte used.[42,49]

The broad peak found in the CV has been suggested to be a result of the nonequivalent electroactive sites in the polymer chain. These nonequivalent sites may arise from several sources. It is known that polypyrrole consists of chains or chain segments with a range of conjugation lengths. Although it has been shown that the average number of pyrrole units[62] in a chain is about 1000, optical,[63] and XPS[64] studies indicate that defects due to β-β linkages can disrupt the conjugation along the backbone and lead to chain segments that average from 5 to 10 pyrrole units in length. The inverse relation between chain length and anodic peak potential is a general feature of conjugated systems and leads to polymer films with a range of formal potentials and thus a broad peak. Heinze et al.[65] performed a study in which thiophene, bithienyl, and quaterthienyl were polymerized over many consecutive potential sweep cycles such that oxidation peaks due to both the monomer and the polymer could be easily separated and identified. By comparing the shape of the CV peaks, they concluded that polymerization of thiophene results in a large number of defects in the form of α-β and β-β linkages, which give rise to twists and bends in the polymer chains that lead to conformational changes during the charging–discharging process. They also concluded that when the conjugation length of the monomer spans at least four units, the number of defects is limited and the oxidation peak of the monomer and polymer doping process becomes indistinguishable as one sharp oxidation peak appears. This weakens the argument that peak broadness is due to different conjugation lengths.

Another feature found in the cyclic voltammograms of polypyrrole that has aroused a great deal of attention in the literature[33,34,36,66–72] is the large background current at potentials greater than the polymer redox peak. Since early observations that this background current is independent of the electrolyte or polarity of the solvent while scaling linearly with the sweep rate (which is indicative of a surface localized process,[33,34] many treatments have attributed the current to capacitive charging of the polymer film. AC capacitance measurements were used to demonstrate that oxidized polypyrrole is effectively a porous metallic electrode with a large surface area and has a double-layer capacitance proportional to the amount of material present.[66] This work also demonstrated that electron transfer reactions can occur at the polypyrrole surface. A theoretical treatment of this problem by Feldberg[67] interpreted the currents in the anodic tail of the CV as purely capacitive because of the large surface area of the polymer.

The Feldberg treatment lead to a picture of a microporous polypyrrole consisting of rods or fibrils with an effective radius of 4 nm and capacitance of 100 F/cm^3. Other work by Feldman, Burgmayer, and Murray[68] found a bulk capacitance of 200 F/cm^3, which would require rods with a radius of 0.5 nm. Thus, 44% of all the pyrrole molecules must be surface species with a uniform density of surface energy states sufficient to allow for the oxidative loss of electrons during the charging process. This leads them to suggest an alternative model involving a less porous microstructure in which the charging sites on the pyrrole are uniformly distributed. Further work by Pickup and Osteryoung[46] summarized these results and concluded that the validity of the two models would depend on the microstructure of polypyrrole when it is swollen with solvent. By examining the electrochemistry in ambient temperature molten salts, they were able to conclude that when swollen with solvent, polypyrrole is nonrigid and contains pores of molecular dimensions. The solvated redox polymer model is more compatible with the observed electrochemical behavior than the porous metal model of Feldberg.

Another approach in describing the origin and nature of the capacitive charging current was taken by Tanguy and Mermilliod.[60,69,70] These researchers used AC impedance and cyclic voltammetry to separate the total electrochemical current into two components. The first is the capacitive current arising from shallowly trapped ions that form an ionic double layer near the polymer chain in the metallic state. These shallowly trapped ions can follow an AC signal and thus produce the capacitive effect. The second is the noncapacitive current arising from deeply trapped ions fixed on the polymer chain and is necessary to ensure electrical conductivity. These deeply trapped ions are released only at low reduction potentials and represent a doping level of 12% to 14%. Other studies dealing with the capacitive charging of polypyrrole,[71,72] as well as some dealing with polyaniline[19] and polythiophene,[12] appear in the literature.

Copolymers and Composites of Conductive Polymers

The formation of copolymers and composites is one of the most useful tools in polymer science in that the physical and mechanical properties of a polymer can be controlled and enhanced. In recent years, this tool has been employed for the study of conducting polymers such as polyacetylene[73] and polypyrrole.[74] These copolymers and composites not only are important because of the resultant mechanical properties, but have also been of great use in the study of conduction mechanisms.

As described earlier, charge transport in conducting polymers is believed to occur through a combination of two primary mechanisms, propagation of charge along the polymer chain and hopping of charge between neighboring chains. The lengths of the conjugated chain segments determine to a large degree whether propagation along the polymer chain or interchain hopping will be the dominant mechanism of charge transport. One method useful for enhancing charge propagation along the chain is lengthening the segments of continuous conjugation. This can be accomplished in two ways. The polymer can be stretched to align the polymer segments, allowing for enhanced order. Stretching of conducting polymers typically results in an increase in conductivity of about one order of magnitude.[75] Another method of ensuring extensive conjugation is to prepare the polymer under circumstances that will result in a low number of defects, followed by orientation. Polyacetylene formed in this manner has been reported to demonstrate conductivities as high as 147,000 ohm^{-1} cm^{-1}, which is close to that of metallic copper and about two orders of magnitude greater than Shirakawa-type polyacetylene.[76]

The preparation of both block[77] and graft[78] copolymers of $(CH)_x$ has been used to impart mechanical integrity and processability to this polymer. At relatively low $(CH)_x$ compositions the block copolymers can be solubilized, allowing their molecular weights to be determined by chromatography. Films of these copolymers contain isolated $(CH)_x$ domains, trapped in insulating matrices, and no significant charge transport is observed. As the composition of $(CH)_x$ is increased, effective processability decreases rapidly with a concurrent improvement in conductivity of the doped film.

Disruption of the conjugation along the polymer backbone can greatly inhibit charge propagation and decreases the ability of the material to conduct. It has been reported[83] that random copolymers of acetylene with methylacetylene can produce conductivities ranging from 10^{-3} to about 400 ohm^{-1} cm^{-1} as the content of methylacetylene is changed. Figure 1.5 shows a decreasing conductivity with methylacetylene content, suggesting that steric interactions along the chain inhibit charge transport.

Similarly, charge propagation along the polymer chain is inhibited in substituted polyheterocycles of which polypyrrole will be used as a typical example. In these systems, charge transport along the chain is sufficiently reduced that interchain hopping becomes the dominant charge transport mechanism.[79]

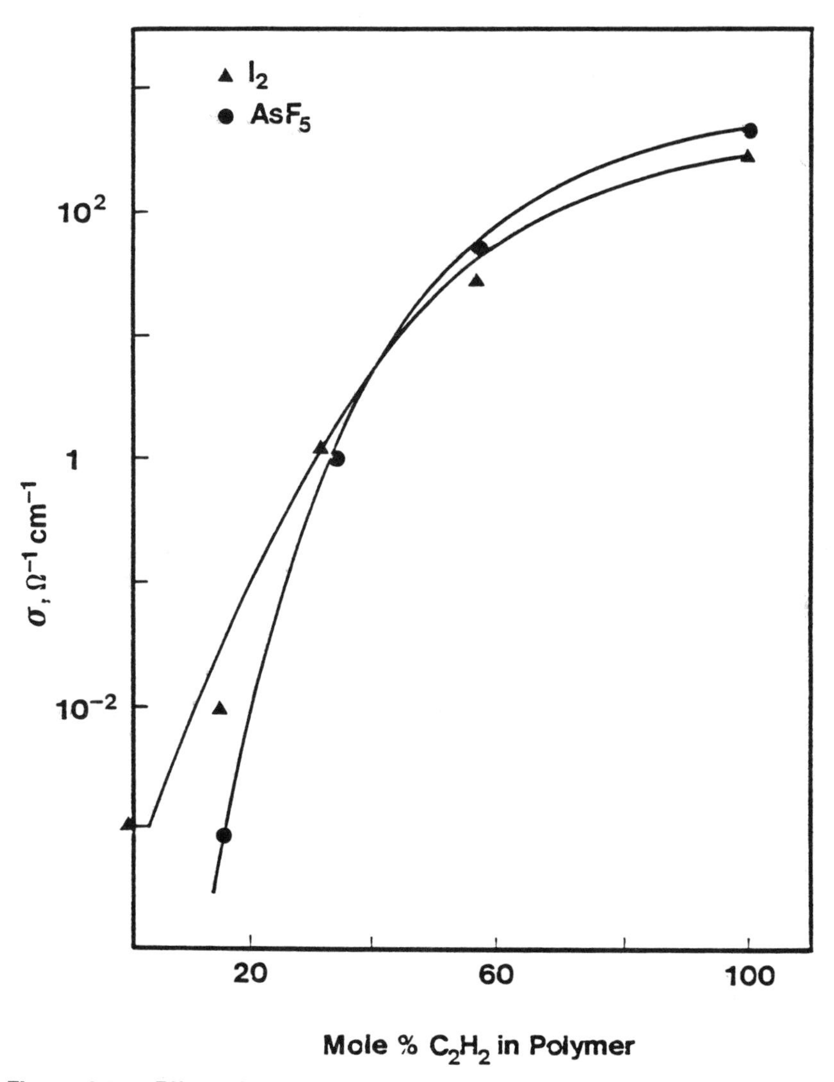

Figure 1.5 Effect of composition on doped conductivity in poly(acetylene-co-methylacetylene).

The formation of composites of conducting polymers in a nonconducting polymer matrix is also used to prepare modified materials. These composites can show practical advantages over the homogeneous materials. They can demonstrate improved environmental stability,[80] mechanical properties,[81] or even good optical transparency.[82] Often the advantages will far outweigh a small decrease in conductivity. In the case of polyacetylene–polyethylene composites, it is found that improvement in the elasticity of the composite material allows improvement of the conductivity through stretching. In fact, the composites containing 40% to 60% polyacetylene by weight were found to attain conductivities as high as 575 ohm^{-1} cm^{-1} when stretched before oxidizing, even higher than those of stretched homogeneous polyacetylene.[83] Recently, $(CH)_x$–polyethylene fibers have been made using the Naarman method with conductivities as high as 6000 ohm^{-1} cm^{-1} with 80% $(C_2H_4)_x$ by weight. It is surprising, however, that composites formed with low contents of polypyrrole actually demonstrate reasonably high conductivities.[84] This allows full advantage to be taken of the mechanical integrity of the matrix polymer. An added benefit encountered with these materials is that they can be formed electrochemically and that polymerization conditions, such as voltage and current, can be used to control the growth of the conducting polymer. Polypyrrole, for instance, can be formed either on the outside surface, on the inside surface, or throughout the insulating polymer matrix.[85]

The complexity of heterogeneous composites makes studies of conduction mechanisms quite difficult. Copolymers, on the other hand, can have a wide composition and allow molecular-level effects to be investigated. Copolymers of pyrrole have been formed in analogous fashion to the acetylene copolymers. In fact, copolymers of pyrrole and *N*-methylpyrrole[86] show much the same dependence of conductivity on composition as the acetylene system as shown in Figure 1.6. A theoretical examination of the structure of this copolymer shows that steric interactions do interfere greatly with conjugation, as shown in Structure 9. Analysis of these insoluble, intractable substances for the amount of incorporated methyl

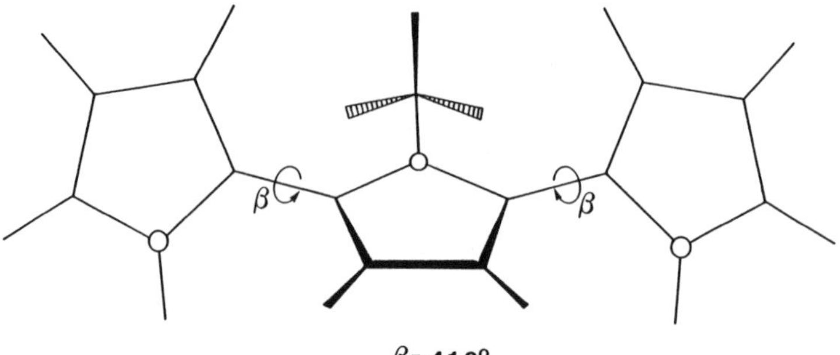

$\beta = 41.8^{\circ}$

9

group is at best inexact because of the limited differences detectable by elemental analysis. Changing the *N*-substituent to a phenyl group proved to be of little help in determining composition, but addition of a bromine

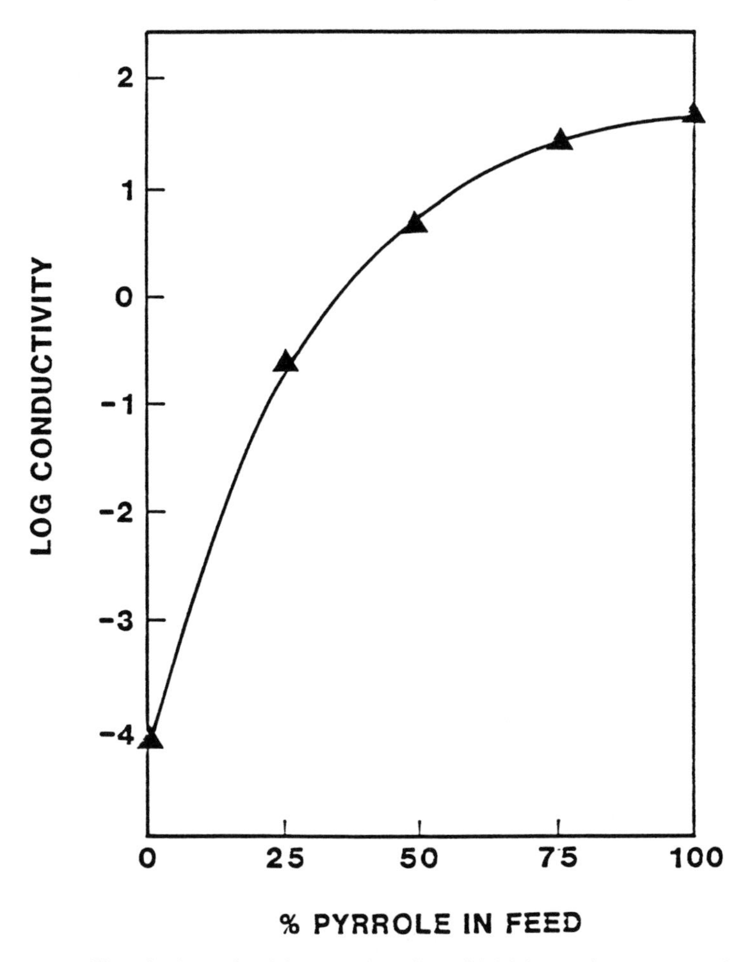

Figure 1.6 Electrical conductivity as a function of initial pyrrole concentration in the feed for poly(pyrrole-co-N-methylpyrrolylium tosylate).

label to the phenyl group allowed elemental analysis to provide direct, accurate analysis of copolymer composition. Analysis of the copolymers revealed that although a compositional dependence of conductivity on feed composition could be determined, analogous to $(CH)_x$ copolymers, as shown in Figure 1.7, the actual dependence of conductivity on composition was found to be linear with composition as shown in Figure 1.8.

Considering copolymers of polypyrrole, one might be led to believe that a lack of conjugation is responsible for the decrease in the conductivity. In fact, the decrease in conductivity can be attributed to a lower dopant level seen by a lower counter-ion content in the polymer, as shown in Figure 1.9. Charges on the polypyrrole backbone are generally

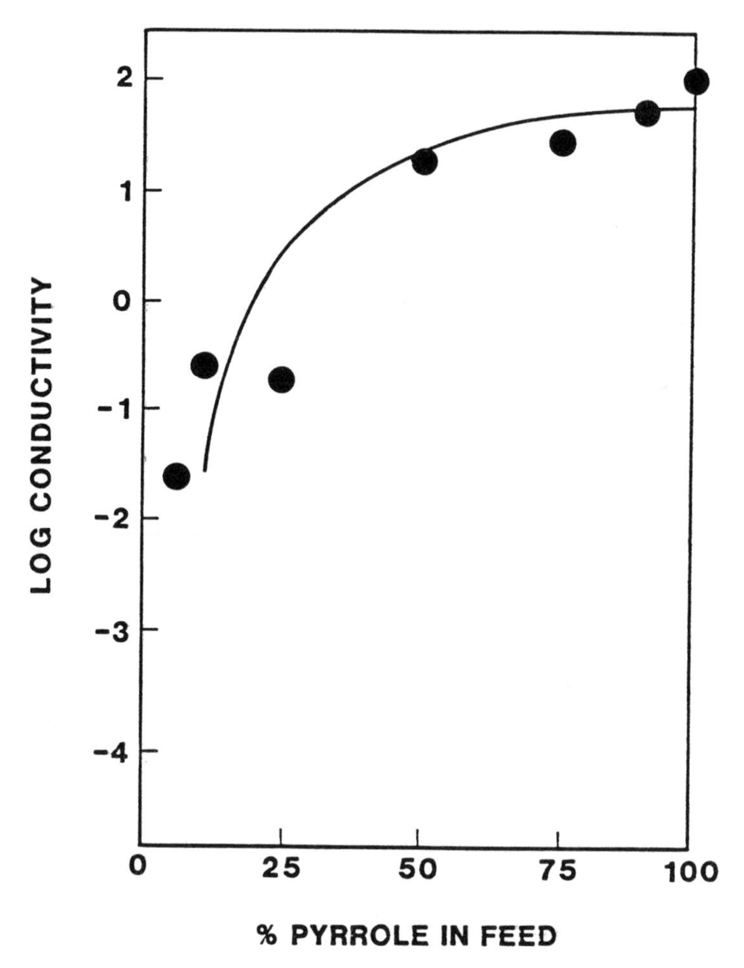

Figure 1.7 Electrical conductivity as a function of initial mol % pyrrole in the feed for poly(pyrrole-co-N-(3-bromophenyl)pyrrolylium tosylate).

believed to be delocalized over approximately four monomeric units.[87] The presence of the substituent decreases the frequency of occurrence of four adjacent, conjugated units. This decreases the number of sites that can support and stabilize a charge carrier. In other words, the substitution inhibits the interchain hopping mechanism of charge transport in polypyrrole and the charge propagation along the polymer chain in polyacetylene.

Another factor that affects the charge transport in these systems is the nature of the counterion. Small nucleophilic counterions appear to bind tightly to the chain, thus limiting the mobility of the charge, and

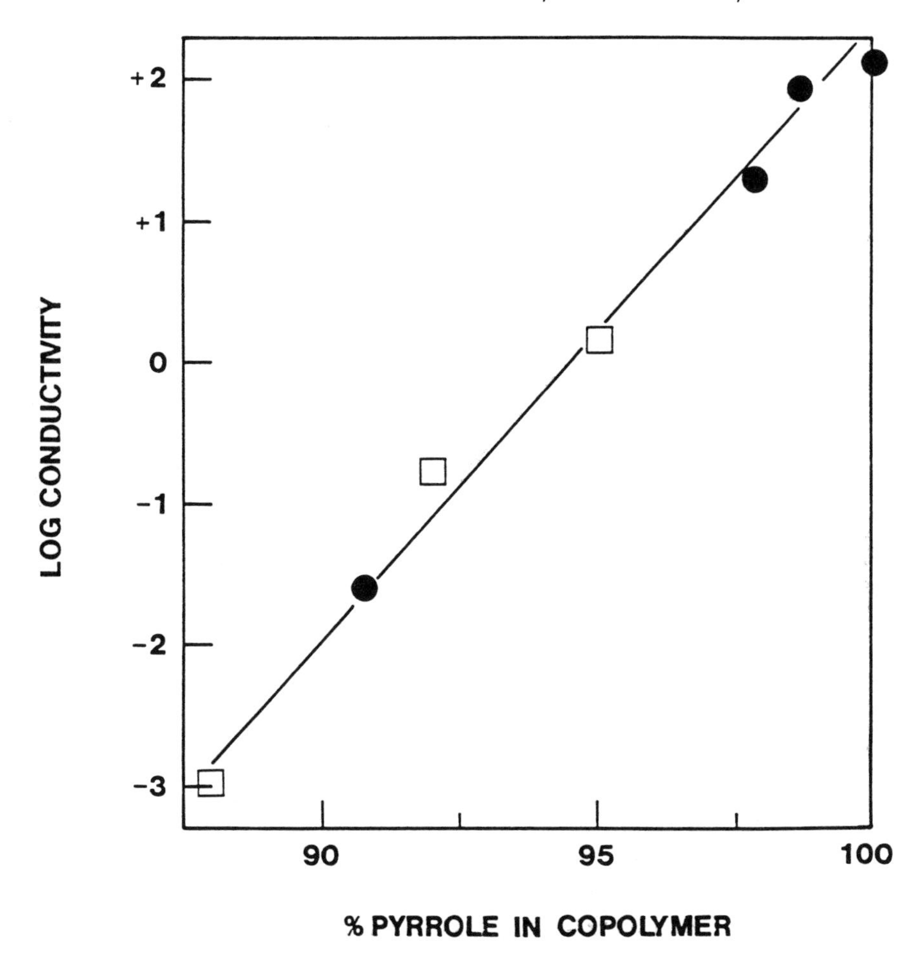

Figure 1.8 Four-probe electrical conductivity as a function of actual mol % pyrrole in the copolymer for poly(pyrrole-co-N-(3-bromophenyl)pyrrolylium tosylate); ○ galvanostatic synthesis; □ potentiostatic synthesis.

larger, delocalized counterions have a more diffuse charge and do not associate highly with the polymer chain.[59] Polymeric counterions have also been employed to immobilize anions in the polymer and allow the mobile species on charging and discharging to be the typically smaller cations.[88–90] The structure of the polymeric counterion in the polymer has also been shown to affect conductivity. If polymer formation and oxidation occur in a solvent that solvates the extended polymer counterion, the conductivity is very different from the same material formed from a solvent in which the counterion is not extended. In fact, the disruption in conductivity is so great that electrochemical

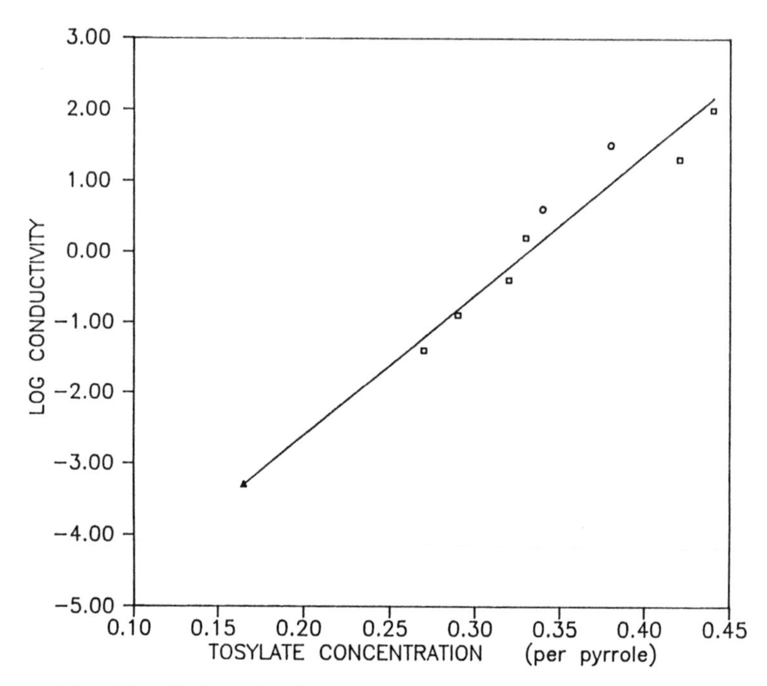

Figure 1.9 Electrical conductivity as a function of dopant ion concentration for poly(pyrrole tosylate) copolymers.

measurements find a percolation model to be most appropriate for describing the conductivity. This is probably due to the inhomogeneities that are introduced into the material.

Processable Conducting Polymers

The most promising conductive polymer should demonstrate good solution or melt processability, in addition to environmental stability, mechanical integrity, and controllable conductivities. As yet, none of the widely known polymers possess all these desired properties, but recent advances in improving the processability of these polymers have been made.

One means of promoting processability in conjugated polymers is by way of soluble precursor polymers. The metathesis polymerization of 7,8-bis(trifluoromethyl)tricyclo[4.2.2.$0^{2,5}$]deca-3,7,9-triene presented earlier (Equation 2) has been developed to provide a soluble precursor system for $(CH)_x$.[8,9] Although this solubility facilitates the isolation, purification, and initial characterization of the $(CH)_x$ system, this precursor is hampered by its thermal instability, that is, even at room temperature it begins to eliminate to form the insoluble conjugated polymer. The metathesis polymerization of 3,6-bis(trifluoromethyl)pentacyclo-[6.2.0.0.^{2,4}.0.^{3,6}.0^{5,7}]dec-9-ene (Equation 6) has been developed handle

(6)

the precursor polymer conveniently at room temperature.[91] In both cases, heating the precursor leads to the elimination of bis(trifluoromethyl)-benzene to form $(CH)_x$.

This soluble precursor method has been used to prepare a number of poly(arylene vinylenes) with a great deal of success, which alleviates problems encountered when either Wittig or dehydrochlorination reactions are used.[4,92] This method proceeds through the polymerization of a bis(dialkylsulphoniumhalide)arylene, outlined in Equation 7. Addition of a base yields a water-soluble polyelectrolyte, poly[arylene dimethy-

$$+ (CH_3)_2S + HCl \qquad (7)$$

lene-α-(dialkylsulphonium halide)], which can be cast into an amorphous polymer film. Heating of the film in an inert atmosphere results in the thermal elimination of dialkyl sulfide and a halogen acid to yield a poly(arylene vinylene) which is insoluble and insulating. Upon exposure to a dopant molecule, the conductivity changes from 10^{-15} ohm^{-1} cm^{-1} to 500 ohm^{-1} cm^{-1}. If the polyelectrolyte film is stretched during the elimination process, the elimination products may serve as plasticizers. This technique has resulted in a highly oriented polymer with a draw ratio of 9 and a doped conductivity as high as 2800 ohm^{-1} cm^{-1}.

Although soluble precursor routes have improved processability, these polymers are still insoluble and unstable to air in the doped conducting state. The combination of improving air stability and concurrent solubility of conducting polymers has been pursued in other systems.

The first discovery of a truly soluble conducting polymer was reported by Frommer et al.[94] Poly(phenylene sulfide) (PPS) was found to be soluble in liquid AsF_3-AsF_5, from which a conducting polymer film could be cast. This procedure presents problems because of the polymer's high environmental reactivity and the toxicity of the solvent used.

Since environmentally stable neutral or oxidized (doped) polymers are obtained from aromatic heterocycles, such as polypyrroles and polythiophenes, much work has been done in attempting to improve their solution properties. It has been found that by placing β-position substituents on polythiophenes, soluble and stable conducting polymers are formed.[95–105] Initially, Elsenbaumer et al.[95] reported a series of 3-alkyl substituted thiophene polymers that were soluble in common organic solvents. These polymers were prepared via a nickel-catalyzed Grignard coupling reaction (Scheme 2) similar to the one used by Kobayashi et

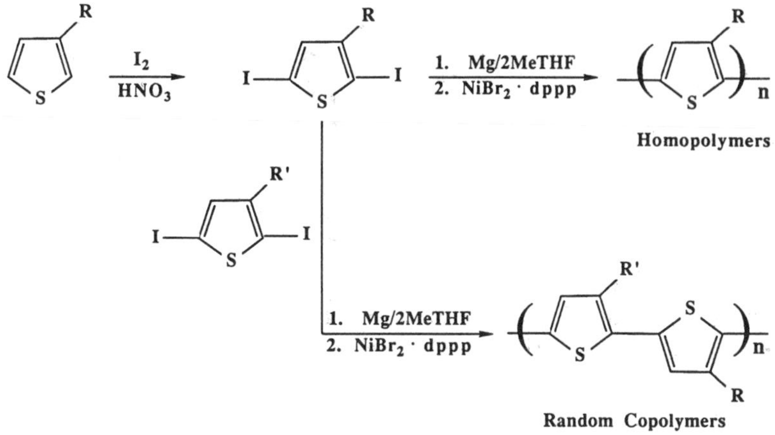

R
I_2
HNO_3
I
S
I
1. Mg/2MeTHF
2. $NiBr_2 \cdot$ dppp
n
Homopolymers
R'
1. Mg/2MeTHF
2. $NiBr_2 \cdot$ dppp
Random Copolymers

Scheme 2

al.[96] to obtain pure neutral polymers. Characterization of these soluble polymers indicated that they have essentially regular linear structures.

The solubility of the homopolymers of 3-alkylthiophenes increases with the size of the side chain (i.e., butyl>ethyl≫methyl). The random copolymers of 3-methylthiophene with 3-butylthiophene or 3-octylthiophene also gives processable materials. The size of the side chains, however, has only a negligible effect on the conductivity of the oxidized polymer.[97]

The polymerization of a disubstituted heterocycle, 3,4-dimethylthiophene,[95] gives a pale yellow polymer. The light color indicates a lower extent of conjugation, possibly resulting from steric interactions forcing adjacent rings to twist out of plane. This polymer is also no longer oxidized by mild oxidizing agents.

Hotta et al.[98] and Sato et al.[99–104] have used another route for preparing these polymers. They electropolymerized the 3-alkyl monomers to obtain polymers in the oxidized conducting form. Solubility was achieved by reducing these polymers to their insulating forms. Similar to the chemically produced polymers, the structural analysis of these polymers indicates a regular 3-alkylthiophene-2,5-diyl structure. The electropolymerization of cyclopenta[c]thiophene, carried out by Roncali et al.,[105] resulted in a polymer with a higher conductivity than the 3,4-dimethylthiophene polymer. Introducing the closed-ring system results in reduction of the steric hindrance and in higher conductivity.

Studies by Hotta et al.[106] on some poly(3-alkylthiophenes), synthesized electrochemically then reduced to the neutral form, demonstrated that these conjugated polymers could be processed from solution and subsequently doped to conducting materials. A weight average molecular weight (M_w) of about 48,000 ($M_w/M_n \approx 2$) was obtained by gel permeation chromatography against polystyrene standards.

The electronic spectra reported by Nowak et al.[107] for reduced poly(3-hexylthiophene) and poly(3-butylthiophene) shows a solid-state interband transition (onset of the $\pi \rightarrow \pi^*$ transition) of about 2.0 eV, similar to that of unsubstituted polythiophene. A band gap blue shift to 2.2 eV was found for these polymers in dilute solution, implying a conformational change of the polymer in solution, presumably to a more disordered form.

Spectral changes in 3-substituted polythiophenes upon chemical (Figure 1.10) or electrochemical oxidation (Figure 1.11) are typical of conducting polymers. In both cases, the depletion of the interband transition is accompanied by the emergence of two subgap absorption[6] upon oxidation. This phenomenon is consistent with predominantly bipolaronic charge storage.[108]

Since these poly(3-alkylthiophenes) are soluble in organic solvents, polymer composites, as described in the preceding section, are possible by solution blending. Conducting polymer composites of poly(3-hexylthiophene) and poly(3-benzylthiophene) in polystyrene, when oxidized with $NOPF_6$, display insulator-metal transitions at the percolation threshold: a volume fraction of $\approx$ 16% conducting polymer in polystyrene.[109] It has also been found that poly(3-alkylthiophenes) with alkyl chain lengths of $n > 4$ are fusible below 300°C.[110] Conducting polymer fibers can thus be made from these polythiophene derivatives by both melt and solution spinning, giving conductivities higher than that of their corresponding films, perhaps because of the alignment of the polymer chains.

Aside from the polymerization of the 3-alkylthiophenes, the polymer of 3-methoxythiophene, obtained by anodic polymerization on carbon surfaces, has been found to be soluble in the reduced state.[111–113] The

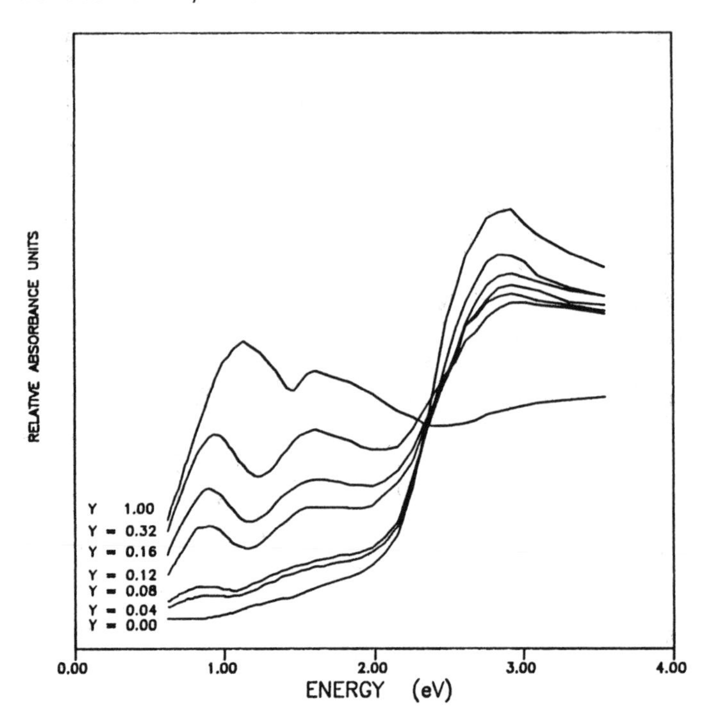

Figure 1.10 Optical spectra for poly(3-ethylmercaptothiophene) during solution oxidation with $NOPF_6$.

reduced polymer is soluble in a variety of solvents, such as DMF, acetonitrile, and benzonitrile. The structural, spectrochemical, and electrochemical studies on this system closely parallel those of 3-alkylthiophene polymers. The oxidation of insulating poly(3-methoxythiophene), electrochemically or with I_2, results in the formation of bipolaronic species. Reduction of the oxidized polymer with three equivalents of triethylamine gave 90% conversion to the insulating polymer. Dopant anion exchange in solution was also studied in this polymer. The replacement of typical anions, such as perchlorate, tetrafluoroborate, or hexafluorophosphate, with other anions that are used as carriers for drug release was analyzed. This processable conductive material has been shown to be applicable for the controlled release of glutamate, a neurotransmitter.

Other β-substituted thiophene polymers have been reported by Bryce et al.[114] Aside from the processability gained from these derivatives, in some cases improved conductivities have been observed. In electrochemically synthesized poly[3-(methoxyethoxyethoxymethyl)thiophene], the polymer is reportedly soluble in the conducting state.

Water-soluble polymers of thiophene derivatives have been obtained

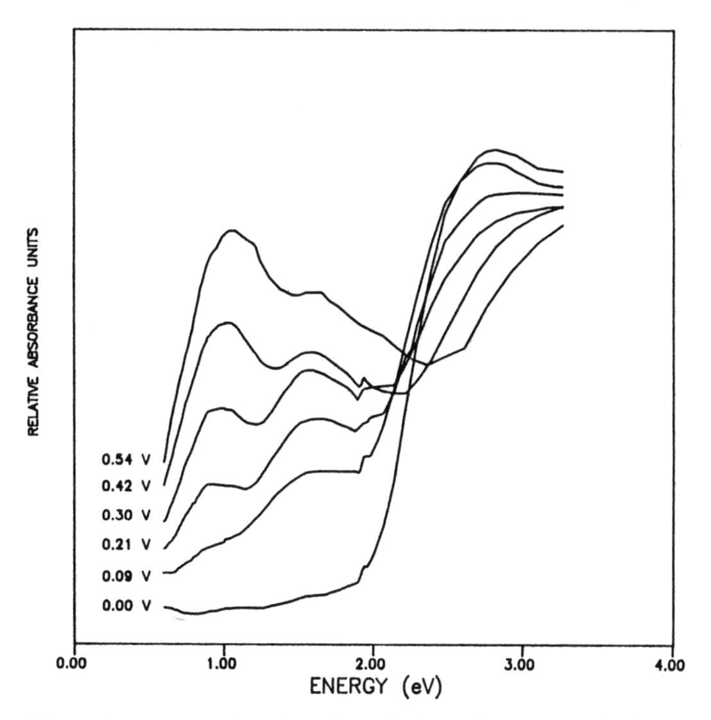

Figure 1.11 Optoelectrochemistry for poly(3-ethylmercaptothiophene).

by using alkylsodium sulfonates as substituents.[115–117] The sodium salts of 3-thiophene-β-ethane sulfonate and 3-thiophene-β-butanesulfonate did not electropolymerize and thus required conversion to their corresponding esters. Conversion to the polymeric sodium salt or the conjugate acid followed polymerization. The polymer of the conjugate acid is interesting, since upon oxidation, it can lose a proton with concurrent electron loss, to form self-doped polymers.[118] These are conducting polymer systems with dopant anions that are attached to the polymer backbone via a covalent bond. The electropolymerization of 3′-propylpotassium sulfonate-2,2′-5′,5″ terthienyl monomer also yields a water-soluble polymer even without using an electrolyte.[115] This gives rise to a self-doped conducting polymer that is water-soluble in the conducting state.

Metal Ion-Containing Polymers

Considerable attention has been focused recently on the synthesis of conducting transition metal-containing, or organometallic, polymers. A number of articles have been published that review the field of organometallic polymers.[119–123] Possible applications for such polymers include high-temperature coatings, electrode and display materials, lasers, and homogeneous catalysts. One of the most interesting areas of metal-containing polymers is that of electrically conducting organometallic polymers. The method of choice for producing conducting organometallic polymers involves complexing transition metals with conjugated bridging ligands. The ability to alter the oxidation state of the metal ion, and thus the charge density along the polymer backbone, provides an alternative route to charge carrier creation as opposed to redox doping. In this manner, intrinsically conductive polymers (polymers that are conductive as synthesized) can be prepared. These systems, either linear or three-dimensional in nature, can be attained with a wide variety of both physical and electronic properties. The structures of these polymers can be represented schematically, as shown in Structure 10. The infinite-chain polymer can also be constructed through direct metal–metal or

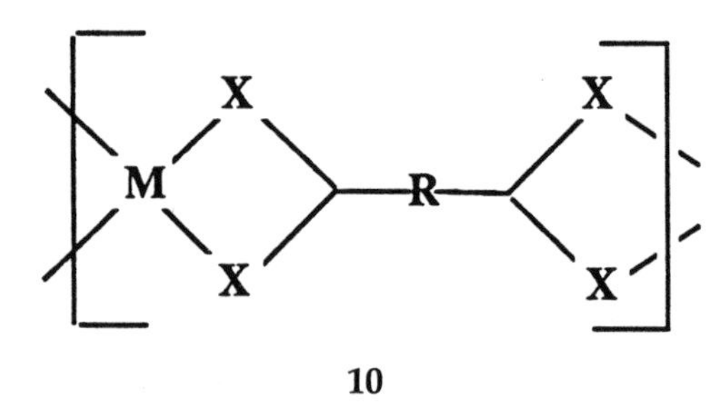

10

metal-bridging ligand–metal interactions, as shown in Structure 11. The following specific systems are representative of these types of organome-

—M—M—M—M—

—M—L—M—L—M—L—M—L—

11

tallic polymers having conductivities ranging from semiconducting (ca. 10^{-6} ohm^{-1} cm^{-1}) to highly conducting (ca. 10^{+3} ohm^{-1} cm^{-1}).

One of the most recently synthesized classes of organometallic conducting polymers consists of the poly(metal tetrathiooxalates).[124–126] These highly conducting linear complexes are synthesized using a 1:1 stoichiometry of the dianionic ligand tetraethylammonium tetrathiooxalate (TEATTO) and divalent salts of Cu, Ni, Pd, Pt, Co, and Fe (Equation 8). These complexes are air-stable, insoluble, and mainly amorphous

$$x\,(Et_4N)_2C_2S_4 + x\,Ni(NO_3)_2 \rightleftarrows [NiC_2S_4]_x + 2x(Et_4N)(NO_3) \quad (8)$$

powders composed of oligomers and containing from three to eight monomer units. The pressed pellet conductivities for these compounds are high, ranging from 3 to 45 ohm^{-1} cm^{-1} for $[NiC_2S_4]_x$ and $[CuC_2S_4]_x$ to 1 ohm^{-1} cm^{-1} for $[PdC_2S_4]_x$. Complexes of TEATTO with Cu^I salts have also been prepared,[126] and conductivities for these complexes are comparable to those obtained for the Cu^{II} salts (1 ohm^{-1} cm^{-1} to 45 ohm^{-1} cm^{-1}). The rigid, ribbonlike structure of these polymers causes them to be insoluble in all solvents. Structural analyses by x-ray scattering methods show the polymers to be composed of metal bisdithiolene-type linkages with some short-range order. These metal tetrathiooxalate complexes are actually *n*-type semiconductors, as evidenced by their low negative thermoelectric power coefficients of ca. $-10\ \mu V\ K^{-1}$. Thermoelectric power measurements and temperature dependence of conductivity studies suggest a variable-range hopping mechanism for conductivity in pellets of these polymers. Vapor phase I_2 oxidation of these complexes leads to a decrease in their electrical conductivity as the concentration of mobile electrons is depleted.

Another example of conducting organometallic polymers is the tetrathiafulvalene-metal bisdithiolene class of compounds. Tetrasodium tetrathiafulvalene tetrathiolate reacts with transition-metal salts (ML_n) to give insoluble, amorphous powders[127–129] having a metal–ligand ratio of approximately 1:1. This reaction results in the formation of insoluble powders that are believed to be oligomers containing a repeating tetrathiafulvalene-metal bisdithiolene unit, as shown in Structure 12. The Ni(II) complex was found to have a conductivity of 30 $ohm^{-1}\ cm^{-1}$ when han-

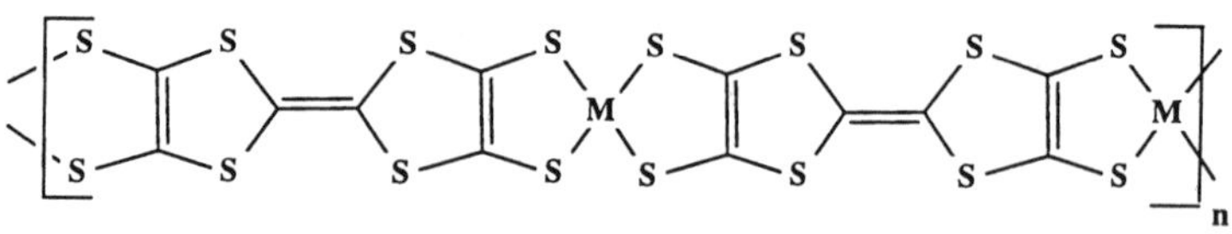

12

dled in air. Conductivities for other transition-metal derivatives were found to be ca. $10^{-1}\ ohm^{-1}\ cm^{-1}$ for Cu(II), $10^{-5}\ ohm^{-1}\ cm^{-1}$ for Fe(II), $10^{-2}\ ohm^{-1}\ cm^{-1}$ for Pt(II), and $10^{-3}\ ohm^{-1}\ cm^{-1}$ for Pd(II). The preceding series of reactions to form the polymeric tetrathiafulvalene–metal bisdithiolenes have also been carried out under inert conditions. These compounds were found to be quasi–one-dimensional conductors. The sulfur atoms in these complexes serve to enhance interstack interactions, thus increasing their conductivity.

Organometallic polymers have also been prepared from the benzene-1,2,4,5-tetrathiolate ligand and divalent transition metals, such as Co, Ni, Fe, and Cu, as shown in Equation 9.[130] Several reasons exist for

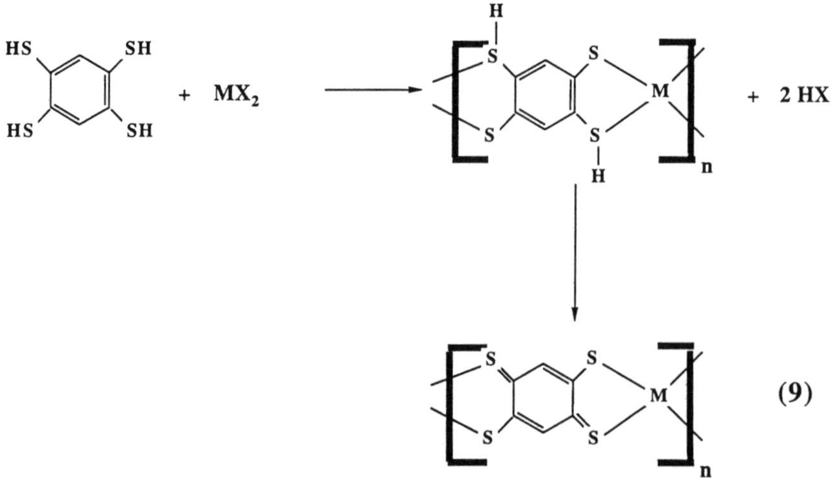

$+\ MX_2 \longrightarrow\ +\ 2\ HX$

(9)

preparing polymers of this type. First, there is the possibility of mixed valence compounds that, along with electron delocalization, will give conducting polymers. The formation of paramagnetic polymers might be possible because of a high density of unpaired spins on the metal atoms. This could lead to ferromagnetic and antiferromagnetic compounds. Finally, the idea of multiple valence states available to the metal atoms suggests the possibility of using these polymers for electrodes in fuel cells.

The polymers obtained are insoluble, amorphous powders, with the exception of the cobalt compound, which is microcrystalline. All the polymers were found to be paramagnetic conductors with conductivities ranging from 10^{-4} to 10^{-1} ohm^{-1} cm^{-1}, and there was no evidence for mixed-valence compounds.

Reaction of the potassium salt of tetrathiosquarate with hexacarbonyl-6A metals[131] gives compounds having bischelated transition metals and the structure shown in Structure 13. Diamagnetic compounds are ob-

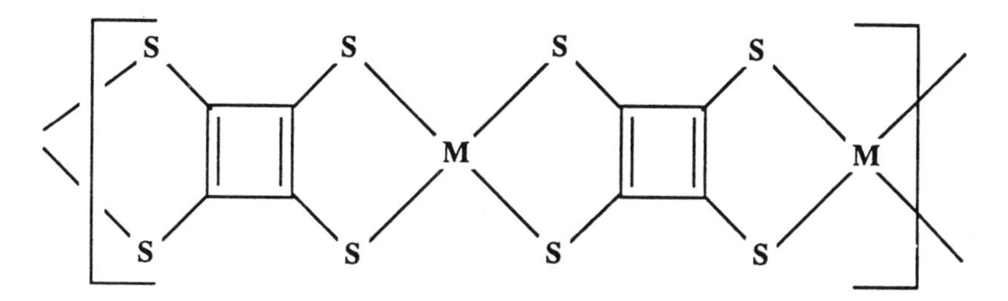

13

tained from the reaction of aqueous solutions of $NiCl_2$ or Na_2PdCl_4 with $K_2C_4S_4$. These Ni- and Pd-based polymers have degrees of polymerization of 10 and 25 and conductivities of 5×10^{-4} ohm^{-1} cm^{-1} and 3×10^{-7} ohm^{-1} cm^{-1}, respectively. The Pt-based polymer can be synthesized by reacting $K_2C_4S_4$ with K_2PtCl_4 or $(PhCN)_2PtCl_2$ to give a blackish-green complex having a 1:1 ratio of $Pt:C_4S_4$ and a conductivity of 6×10^{-7} ohm^{-1} cm^{-1}.

The electronic structure of these complexes has been studied theoretically.[132,133] The interchain interactions are found to lead to an equivalency of the double and single bonds in the C_4S_4 moiety.

Upon comparison of these complexes with other 1,2-dithiolene compounds, several differences are obvious: (1) no reversible redox properties are observed for the squarate complexes; (2) tetrahedral structures seem to be common in squarate complexes, unlike compounds of 1,2-dithiolenes; (3) low oxidation states of the metal are more favored in the squarates. These differences arise from the unusually large S–S intraligand distance in the squarate complexes.

Recently we have prepared polymers in our laboratory by reacting nickel bromide with a tetraanion derived from bis(4-styryldithio-carbonate)oxide (Equation 10).[134] The resulting compound, poly [bis(4-oxystyryl-dithiolene)nickel], is *soluble* in its reduced form. The compound oxidizes upon exposure to air to give a partially soluble to insoluble powder. A conductivity of ca. 10^{-6} ohm^{-1} cm^{-1} has been measured for the oxidized polymer. The complex shows a broad absorption in the VIS–NIR spectrum at 930 nm, as is seen for other metal bisdithiolene complexes. This suggests its possible use as a near-IR dye for optical storage and laser applications. An interesting aspect of these soluble metal complex polymers is that they are electrochemically active in solution. A cyclic voltammetric analysis shows that the as synthesized $[NiL_2]^{2-}$ form can be sequentially, and reversibly, switched to the $[NiL_2]^{1-}$ and $[NiL_2]^0$ oxidation states.

As detailed, a number of transition-metal-containing polymers having high conductivities have been synthesized to date. Though the physical properties of many of these compounds are poor, their novel electronic properties suggest they may ultimately provide new materials of interest. Future research should focus on the synthesis of soluble organometallic polymers, such as the poly[bis(4-oxystyryldithiolene)nickel] compound. The solubility of this polymer will facilitate studies of structural, physical, and electronic properties that are difficult on the previously studied insoluble, amorphous metal-containing polymers.

Present and Future Applications of Conductive Polymers

The presence of a large number of conductive polymers with such a wide range of physical and electronic properties has led to a number of potential applications. The development of some of these applications to commercial viability is just now beginning. The ability to tailor the electrical properties of these systems is one of their most attractive features and, coupled with improved stability and processability relative to the original conductive polymer systems, new applications are surely on the horizon.

An extensive discussion of all the proposed applications for conductive polymers would be quite lengthy. The scope of these applications includes electrodes for batteries, fuel cells and capacitors, electrochromics, chemical and biochemical sensors, EMI and power cable shielding, ion exchange and release devices, and neutron detection. A few select areas have been chosen for description here.

Remotely Readable Indicators

One program, being carried out at Allied-Signal, Inc., has shown that electrically conductive polymers can be used as ambient-responsive elements for remotely readable indicator devices. These indicators can detect a number of diverse and changing conditions, including time–temperature, temperature limit, humidity, radiation dosage, mechanical abuse, and chemical release. They utilize the ability of conjugated polymers to change their electrical properties during reaction with various redox agents (dopants) or via their instability to moisture and heat.

Many of these indicators are based on technology utilized in radio frequency (RF) (Checkpoint Systems, Inc., Thorofare, NJ) and antitheft targets. They are low cost and work by the interaction of a small RF resonance circuit with a source and receiver. The circuit absorbs and remits RF energy, which varies in intensity with a scanning transmitter's instantaneous frequency. Monitoring the frequency-dependent output, a selective response attributable to the circuit develops and is accumulated in memory, which provides the detector response. When the circuit is covered by a conductor, such as an aluminum foil, the RF does not detect the resonator and the product produces no response.

Conducting polymers can be applied as overlayers on these circuit resonators, and thus the resonator's response will be controlled by the polymer's electrical properties. The devices use the ability of these polymers to change their electrical properties significantly via a specific environmental exposure. This may include

1. Large conductivity increases upon exposure to oxidants or reductants (i.e., *p*- or *n*-type dopants).
2. Large conductivity decreases upon compensation of these dopants by suitable reagents (such as amines).
3. Induced degradation processes, such as a conductive complex that is unstable to water being exposed to air or thermal degradation.

These major electrical property changes reverse the response of the RF circuit underneath. The circuit can then be read externally to indicate the specific history of a particular package. The preceding actions can be caused to occur by a number of means. For example, the response of a time–temperature indicator might be controlled by

1. The thermal degradation of a doped polymer.
2. The diffusion of a dopant or compensating agent at a predetermined rate through a barrier prior to contacting the reactive polymer.
3. The thermal generation of a dopant or compensating agent that can interact with exposed polymer.

What in many cases is considered a drawback of these conductive polymer systems (i.e., their high reactivity with concomitant change in electrical properties—both increases and decreases) proves to be a benefit here.

Biosensors

Resistance changes that occur in conducting polymers when exposed to oxidants in solution can be used in specific sensor applications. One example of a series of biosensors, under development at Ohmnicron Corporation (Pennington, NJ), utilizes the ability of triiodide to oxidize polyacetylene as a means of measuring the concentration of glucose in solution. The technique, shown in Figure 1.12, begins with the glucose oxidase–mediated oxidation of glucose in solution. This creates hydrogen peroxide, which in the presence of catalytic lactoperoxidase oxidizes

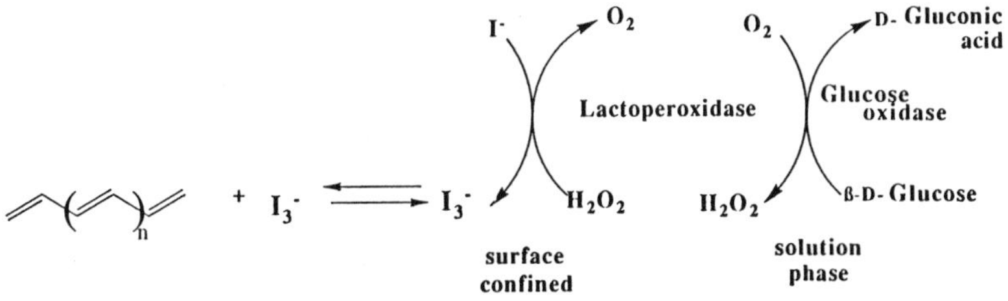

Figure 1.12 Application of $(CH)_x$ to a glucose concentration measurement system.

iodide in buffer solution to triiodide. The triiodide, being a good dopant for polyacetylene, oxidizes the polymer, causing a resistance change that is proportional to the concentration of glucose in solution. The glucose content is actually determined as a percent conductivity change in a specified amount of time and changes linearly with glucose concentration. This and other biosensor concepts are being developed at Ohmnicron. Sensitivity enhancements have been obtained using surface-confined enzyme systems, and new solid-phase enzyme immunoassays (including *Streptococcus A* and *Salmonella*) are presently being pursued.

Storage Battery and Fuel Cell Electrodes

Probably the most publicized and promising of the current applications, rechargeable storage batteries using conducting polymer electrode materials, are getting closer to commercialization. Progress at both Allied-Signal, Inc. (Morristown, NJ) and BASF Germany in cooperation with VARTA Batterie AG has led to the development of prototype cells with characteristics comparable to, and in some instances better than, nickel–cadmium cells now on the market.

The mechanisms of charge storage and discharge in conducting polymers have come under a significant amount of scrutiny as researchers investigate the basic phenomena behind these processes and attempt to increase such parameters as cell voltage, energy density, and power output. The redox doping reactions used to charge and discharge these electrode materials are inherently different from those used in conventional batteries because the electrode is not dissolved and redeposited. Instead a conducting polymer battery, such as a polypyrrole–lithium cell, operates by the oxidation and reduction of the conjugated polymer backbone, the doping reactions discussed earlier.

During the charging portion of the cycle a positive potential is applied to the polypyrrole electrode relative to the lithium electrode. As the polypyrrole oxidizes, anions in the electrolyte enter the porous polymer to balance the charge created. Simultaneously, lithium ions in electrolyte are electrodeposited at the lithium surface. When this charged cell, which has an open cell voltage on the order of 3 V, is connected to a load, current spontaneously runs through the external circuit. Electrons are removed from the lithium, causing lithium ions to reenter the electrolyte and to pass through the load and into the oxidized polymer. The positive sites on the polymer are reduced, releasing the charge-balancing anions back to electrolyte. This process can be repeated for many cycles as required for a typical secondary battery cell.

Fuel cells[138] can be constructed using conductive polymers as electrocatalytic materials. This is illustrated for the case of $(C_2H_4)_x$ in which oxidation of the polymer film in the presence of HBF_4 leads to a conductive $(CH)_x$ with BF_4^- dopant anions as shown in Equation 11. Other

$$4\ (CH)_x + 4xy\ HBF_4 + xy\ O_2 \longrightarrow 4\ [CH^{y+}\ (BF_4)_y^-]_x + 2xy\ H_2O \qquad (11)$$

reagents, including perchloric acid and benzoquinone, have been used to oxidize $(CH)_x$ to the highly conducting state in aqueous solutions. Electrically connecting a $[CH^{y+}(BF_4)_{y-}]_x$ electrode with a Pb electrode in a 7.4 *M* aqueous HBF_4 solution leads to a spontaneous electrochemical reaction in which the polymer is reduced to the neutral state, emitting BF_{4-} into the electrolyte, and the Pb is oxidized to Pb^{2+}. This reaction can then be used to construct a fuel cell using O_2 as the fuel material. If during the discharge of the $[(CH)^{y+}(BF_4)_{y+}]_x \mid Pb(BF_4)_2 \mid Pb^0$ cell, O_2 is introduced, the electromechanically reduced polymer will be *simultaneously* chemically oxidized with continuous production of electricity. In this manner, the conducting polymer will behave catalytically, undergoing no net chemical change, and the Pb electrode and oxidizing agent will be converted to products. Research into potential "fuel" materials may yield a system by means of which conversion to electricity in this manner is economically viable.

Conclusion

A wide range of electrically conducting and electroactive polymer systems is now available. Recent developments of materials having improved processability and ambient stability relative to the early systems are leading the way as these materials approach commercial viability. These materials will be useful when the unique properties of a conductive polymer—such as controllable conductivity, elasticity in composites, electrochromic behavior, and chemical reactivity in sensor applications—are important. Continued research with an interdisciplinary approach, which includes chemistry, physics, materials science, and device development, will surely lead to exciting discoveries in the years to come.

References

[1] Ito, T., Shirakawa, H., Ikeda, *S. J. Polym. Sci., Polym. Chem.* 1974, *12*, 11.

[2] Chien, J. C. W., *Polyacetylene: Chemistry, Physics and Materials Science*. Orlando, Fla.: Academic Press, 1984. Skotheim, T., ed., *Handbook of Conducting Polymers*. New York: Marcel Dekker, 1986.

[3] Ivory, D. M., Miller, G. G., Sowa, J. M., Shacklette, L. W., Chance, R. R., Baughman, R. H., *J. Chem. Phys.* 1979, *71*, 1506.

[4] Wnek, G. E., Chien, J. C. W., Karasz, F. E., Lillya, C. P., *Polym Comm.* 1979, *20*, 143.

[5] Rabolt, J. R., Clarke, T. C., Kanazawa, K. K., Reynolds, J. R., Street, G. B. J., *Chem. Soc. Chem. Comm.* 1980, 347.

[6] Diaz, A. F., Kanazawa, K. K., Gardini, G. P., *J. Chem. Soc. Chem. Commun.* 1979, 635.

[7] Tourillon, G. T., Garnier, F., *J. Electroanal. Chem.* 1982, *135*, 173–178.

[8] Edwards, J. H., Feast, W. J., *Polymer* 1980, *21*, 595.

[9] Edwards, J. H., Feast, W. J., Bott, D. C., *Polymer* 1984, *25*, 395.

[10] Dall'olio, A., Dascola, Y., Varacca, V., Bocchi, V., *Compt. Rend.* 1968, *C267*, 433.

[11] Kanazawa, K. K., Diaz, A. F., Geiss, R. H., Gill, W. D., Kwak, J. F., Rabolt, J. F., Street, G. B., *J. Chem. Soc. Chem. Commun.* 1979, 854–855.

[12] Waltman, R. S., Bargon, J., Diaz, A. F., *J. Phys. Chem.* 1983, *87*, 1459–1463.

[13] Yamamoto, T., Sanechiraand, K., Yamaamoto, A., *J. Polym. Sci. Polym. Lett. Ed.* 1980, *18*, 9.

[14] Tourillon, G., Garnier, F., *J. Electrochem. Soc.* 1983, *130*, 2042–2044.

[15] Ambrose, J. F., Nelson, R. F., *J. Electrochem. Soc.* 1968, *115*, 1159.

[16] Bargon, J., Mohmand, M., Waltman, R. J., *IBM J. Res. Develop.* 1987, *27*, 330.

[17] Bargon, J., Mohmand, S., Waltman, R. J., *Mol. Cryst. Liq. Cryst.* 1983, *93*, 279.

[18] MacDiarmid, A. G., Chiang, M. C., Halpern, M., Huang, W. S., Krawczyk, J. R., Mammone, R. J., Mu, S. L., Somasiri, N. L. D., Wu, W., *Polym. Prepr. Am. Chem. Soc. Div. Polym. Chem.* 1984, *25*, 248.

[19] Kobayashi, T., Yoneyama, H., Tamura, H., *J. Electroanal. Chem.* 1984, *177*, 281.

[20] Sasaki, K., Kaya, M., Yawo, J., Kitani, A., Kunai, K., *J. Electroanal. Chem.* 1981, *215*, 401–407.

[21] Rubenstein, I., Sabatani, E., Rashpon, J. R., *J. ElectroChem. Soc.* 1987, *134*, 3078–3083.

[22] Yoshino, K., Kohno, Y., Shiraisni, T., Kaneto, K., Inoue, S., Tsukagoshi, K., *Synth. Met.* 1985, *10*, 319.

[23] Lewis, S. G., Ginley, P. S., Frank, A. J., *J. Appl. Phys.* 1987, *62*, 190–194.

[24] Dian, G., Merlet, N., Barbey, G., Outurquin, F., Paulmier, G., *J. Electroanal. Chem.* 1987, *238*, 225–237.

[25] Hay, A. S., Blanchard, H. S., Enders, G. I., Eustance, J. W., *J. Am. Chem. Soc.* 1959, *81*, 6335.

[26] Mengoli, G., Musiani, M. M., *J. Electrochem. Soc.* 1987, *134*, 6436.

[27] Suchira, E. T., Nishide, H., Yamamoto, K., Oshida, S., *Macromolecules* 1987, *20*, 2315–2316.

[28] Said, M., Tanaka, S., Kaeriyama, K., *J. Chem. Soc. Chem. Commun.* 1985, 713.

[29] Roncali, J., Garnier, F., *Nouv. J. Chim.* 1986, *10*, 237.

[30] Garnier, F., Tourillon, G., Gazard, M., DuBois, J. C., *J. Electroanal. Chem.* 1983, *148*, 299.

[31] Waltman, R. J., Diaz, A. F., Bargon, J., *J. Electrochem. Soc.* 1985, *132*, 831.

[32] Waltman, R. J., Diaz, A. F., Bargon, J., *J. Electrochem. Soc.* 1984, *131*, 1452.

[33] Diaz, A., *Chemica Scripta* 1981, *17*, 145.

[34] Kanazawa, K. K., Diaz, A. F., Krounbi, M. T., Street, G. B. *Synth. Met.* 1981, *4*, 119.

[35] Asavapiriyanont, S., Chandler, G. K., Gunawardena, G. A., Pletcher, D., *J. Electroanal. Chem.* 1984, *177*, 229.

[36] Asavapiriyanont, S., Chandler, G. K., Gunawardena, G. A., Pletcher, D., *J. Electroanal. Chem.* 1984, *177*, 245.

[37] Hillman, A. R., Mallen, E. F., *J. Electroanal. Chem.* 1987, *220*, 351.

[38] Downard, A. J. Pletcher, D., *J. Electroanal. Chem.* 1986, *206*, 139.

[39] Downard, A. J., Pletcher, D., *J. Electroanal. Chem.* 1986, *206*, 147.

[40] Marcos, H. L., Rodriguez, I., Gonzalez-Velasco, J., *Electrochim. Acta* 1987, *32*, 1453.

[41] Noftle, R. E., Pletcher, D., *J. Electroanal. Chem.* 1987, *227*, 229.

[42] Baker, C. K., Reynolds, J. R., *Polym. Preprints* 1987, *28*, 284.

[43] Sluyters-Rehbach, M., Wisenberg, J. H. O. J., Bosco, E., Sluyters, J. H., *J. Electroanal. Chem.* 1987, *235*, 259.

[44] Genies, E. M., Bidan, G., Diaz, A. F., *J. Electroanal. Chem.* 1983, *149*, 101.

[45] Zotti, G., Cattarin, C., Comisso, N., *J. Electroanal. Chem.* 1987, *235*, 259.

[46] Pickup, P. G., Osteryoung, R. A., *J. Am. Chem. Soc.* 1984, *106*, 2294.

[47] Funt, B. L., Lowen, S. V., *Synth. Met.* 1985, *11*, 129.

[48] Rodriquez, I., Marcos, M. L., Gonzalez-Velasco, J., *Electrochim. Acta* 1987, *32*, 1181.

[49] Baker, C. K., Reynolds, J. R., *J. Electroanal. Chem.*, submitted.

[50] Poropatic, P. A., Toyooka, R., Reynolds, J. R., unpublished results.

[51] Diaz, A. F., Crowly, J. I., Bargon, J., Gardini, G. P., Torrance, J. B., *J. Electroanal. Chem.* 1981, *121*, 355.

[52] Heinze, J., Mortensen, J., Müllen, K., Andrainer, S., *J. Chem. Soc. Chem. Commun.* 1987, 701.

[53] Brédas, J. L., Silbey, R., Boudreauz, D. S., Chance, R. R., *J. Am. Chem. Soc.* 1983, *103*, 6555.

[54] Wynne, K. J., Street, G. B., *Macromolecules* 1985, *18*, 2361.

[55] Warren, L. F., Anderson, D. P., *J. Electrochem. Soc.* 1987, *134*, 101.

[56] Salmon, M., Diaz, A. F., Logan, A. J., Krounbi, M., Bargon, J., *J. Mol. Cryst. Liq. Cryst.* 1982, *83*, 265.

[57] Diaz, A. F., Castillo, J. I., Logan, J. A., Lee, W. Y., *J. Electroanal. Chem.* 1981, *129*, 115.

[58] Marque, P., Roncali, J., Garnier, F., *J. Electroanal. Chem.* 1987, *218*, 107.

[59] Shimidzu, T., Ohtani, A., Iyoda, T., Honda, K., *J. Electroanal. Chem.* 1987, *224*, 123.

[60] Tanguy, J., Mermilliod, N., *Synth. Met.* 1987, *21*, 129.

[61] Diaz, A. F., Castillo, J. J., *J. Chem. Soc. Chem. Commun.* 1980, 397.

[62] Nazzal, A., Street, G. B., *J. Chem. Soc. Chem. Commun.* 1984, 83.

[63] Yakashi, K., Lauchlan, L. J., Clarke, T. C., Street, G. B., *J. Chem. Phys.* 1983, *79*, 4774.

[64] Pfluger, P., Steet, G. B., *J. Physiq.* 1983, *44*, C3–609.

[65] Heinze, J., Mortensen, J., Hinkelmann, K., *Synth. Met.* 1987, *21*, 209.

[66] Bull, R. A., Fan, F. F., Bard, A. J., *J. Electrochem. Soc.* 1982, *129*, 1009.

[67] Feldberg, S. W., *J. Am. Chem. Soc.* 1984, *106*, 4671.

[68] Feldman, B. J., Burgmayer, P., Murray, R. W., *J. Am. Chem. Soc.* 1985, *107*, 872.

[69] Tanguy, J., Mermilliod, N., Hoclet, M., *J. Electrochem. Soc.* 1987, *134*, 795.

[70] Tanguy, J., Mermilliod, N., Hoclet, M., *Intl. Conf. Synth. Met.* 1986, Kyoto, Japan.

[71] Oudard, J. F., Vieil, E., *Ann. Physiq.* 1986, *11*, 17.

[72] Jakobs, R. C. M., Janssen, L. J. J., Barendrecht, E., *Chem. Reci. Soc. J.R.*, Netherlands, 1984, 103.

[73] Chien, J. C. W., Wnek, G. E., Karasz, F. E., Hirsch, J. A., *Macromolecules* 1985, *23*, 1687.

[74] Reynolds, J. R. Poropatic, P. A., Toyooka, R. L., *Macromolecules* 1987, *20*, 1184.

[75] Hinooka, M., Doi, T., *Synthetic Metals* 1987, *17*, 209.

[76] Baseseu, N., Liu, Z.-X., Moses, D., Heeger, A. J., Naarman, H., Theophilou, N., *Nature* 1987, *327*, 403.

[77] Galvin, M. E., Wnek, G. E., *Polymer Bull.* 1985, *13*, 109.

[78] Destri, S., Catellani, M., Botognesi, A., *Macromol. Chem. Rapid. Commun.* 1984, *5*, 353.

[79] Bunhs, E., Hodge, I. M., *J. Chem. Phys.* 1985, *83*, 5976.

[80] DePaoli, M., Waltman, R. J., Dinzand, A. F., Bargon, J., *J. Polym. Sci. Polym. Chem. Ed.* 1985, *23*, 1687.

[81] DePaoli, M., Waltman, R. J., Dinzand, A. F., Bargon, J., *J. Chem. Soc. Chem. Commun.* 1984, 1015.

[82] Niwa, O., Hihito, M., Tamamura, T., *Appl. Phys. Lett.* 1985, *46*, 795.

[83] Galvin, M. E., Wnek, G. E., *Polymer Commun.* 1982, *23*, 795.

[84] Wang, T. T., Tasaka, S., Hutton, R. S., Lu, P. Y., *J. Chem. Soc. Chem. Commun.* 1985, 1343.

[85] Lindsay, S. E., Street, G. B., *Synth. Met.* 1984, *10*, 67.

[86] Reynolds, J. R., Poropatic, P. A., Toyooka, R. L., *Synth. Met.* 1987, *18*, 95.

[87] Reynolds, J. R., *J. Molec. Elec.* 1986, *2*, 1.

[88] Glutzhofer, D. T., Ulanshi, J., Wegnea, G., *Polymer* 1987, *28*, 449.

[89] Shigehara, K., Hara, M., Yamada, A., *Synth. Met.* 1987, *18*, 721.

[90] Novak, P., *J. Power Sources* 1987, *21*, 17.

[91] Feast, W. J., and Winter, J. N., *J. Chem. Soc. Chem. Comm.* 1985, 202.

[92] Gourley, K. D., Lillya, C. P., Reynolds, J. R., Chien, J. C. W., *Macromolecules* 1984, *17*, 1025.

[93] Wessling, R. A., Zimmerman, R. G., U.S. Patent 3,401,152 (1968); Wessling, R. A., Zimmerman, R. G., U.S. Patent 3,706,677 (1972).

[94] Frommer, J., Elsenbaumer, R. L., Chance, R. R., in T. Davidson, ed., *ACS Symposium Series 242, Polymers in Electronics*, American Chemical Society, Washington, D.C., 1984, 447.

[95] Elsenbaumer, R. L., Jen, K. Y., Oboodi, R., *Synth. Met.* 1986, *15*, 169.

[96] Kobayashi, M., Chen, J., Moraes, T. C., Heeger, A. J., Wudl, F., *Synth. Met.* 1984, *9*, 77.

[97] Elsenbaumer, R. L., Jen, K. Y., Miller, G. G., Shacklette, L. W., *Synth. Met.* 1987, *18*, 277; Jen, K. Y., Oboodi, R., Elsenbaumer, R. L., *Polymeric Materials: Science and Engineering, Proceedings of the ACS Division of Polymeric Materials, Science and Engineering* 1985, *53*, 79.

[98] Hotta, S., Hosaka, T., Soga, M., Shimotsuma, W., *Synth. Met.* 1984, *9*, 381.

[99] Sato, M., Tanaka, S., Kaeriyama, K., *J. Chem. Soc. Chem. Comm.* 1986, 873.

[100] Kaeriyama, K., Sato, M., Tanaka, S., *Synth. Met.* 1987, *18*, 233.

[101] Sato, M., Tanaka, S., Kaeriyama, K., *Makromol. Chem.* 1987, *188*, 1763.

[102] Sato, M., Tanaka, S., Kaeriyama, K., *J. Chem. Soc. Chem. Comm.* 1985, 713.

[103] Sato, M., Tanaka, S., Kaeriyama, K., *Synth. Met.* 1986, *14*, 279.

[104] Sato, M., Tanaka, S., Kaeriyama, K., *J. Chem. Soc. Chem. Commun.* 1987, 172.

[105] Roncali, J., Garnier, F., Garreau, R., Lemaire, M., *J. Chem. Soc. Chem. Commun.* 1987, 1500.

[106] Hotta, S., Rughooputh, S. D. D. V., Heeger, A. J., Wudl, F., *Macromolecules* 1987, *20*, 212.

[107] Nowak, M. J., Rughooputh, S. D. D. V., Hotta, S., Heeger, A. J., *Macromolecules* 1987, *20*, 965.

[108] Brédas, J. L., Street, G. B., *Acc. Chem. Res.* 1985, *18*, 309; Brédas, J. L., Themans, B., Fripiat, J. G., Andre, J. M., Chance, R. R., *Phys. Rev. B* 1984, *29*, 6761; Brédas, J. L., Scott, J. C., Yakushi, K., Street, G. B., *Phys. Rev. B* 1984, *30*, 1023.

[109] Hotta, S., Rughooputh, S. D. D. V., Heeger, A. J., *Synth. Met.* 1987, *22*, 79.

[110] Yoshino, K., Nakajima, M., Fujii, M., Sugimoto, R., *Polym. Commun.* 1987, *28*, 309.

[111] Blankespoor, R. L., Miller, L. L., *J. Chem. Soc. Chem. Commun.* 1985, 90.

[112] Chang, A. C., Blankespoor, R. L., Miller, L. L., *J. Electroanal. Chem.* 1987, *236*, 239.

[113] Chang, A. C., Miller, L. L., *Synth. Met.* 1987, *22*, 71.

[114] Bryce, M. R., Chissel, A., Kathigamanathan, P., Parker, D., Smith, N. R. M., *J. Chem. Soc. Chem. Commun.* 1987, 466.

[115] Havinga, E. E., van Horssen, L. W., ten Hoeve, W., Wynberg, Meijer, E. W., *Polym. Bull.* 1987, *18*, 277.

[116] Patil, A. O., Ikenone, Y., Wudl, F., Heeger, A. J., *J. Am. Chem. Soc.* 1987, *109*, 1858.

[117] Patil, A. O., Ikenone, Y., Basescu, N., Colaneri, N., Chen, J., Wudl, F., Heeger, A. J., *Synth. Met.* 1987, *20*, 151.

[118] Frommer, J., *Acc. Chem. Res.* 1986, *19*, 2.

[119] Sheats, J. E., Pittman, C. U., Carraher, C. E., *Chem. Brit.* 1984, 709.

[120] Carraher, C. E., *J. Chem. Ed.* 1981, *58*, 921.

[121] Wohrle, D., *Adv. Polym. Sci.* 1983, *50*, 45.

[122] Miller, J. S., ed., *Extended Linear Chain Compounds*, New York: Plenum, 1982, Vol. 1–3.

[123] Pittman, C. U., Jr., Carraher, C. E., Jr., Reynolds, J. R., in *Mark-Bikales-Overberger-Menges: Encyclopedia of Polymer Science and Engineering*, Vol. 10, New York: Wiley, 1987, pp. 541–594.

[124] Reynolds, J. R., Karasz, F. E., Lillya, C. P., Chien, J. C. W., *J. Chem. Soc. Chem. Commun.* 1985, 268.

[125] Reynolds, J. R., Lillya, C. P., Chien, J. C. W., *Macromolecules* 1987, *20*, 1184.

[126] Reynolds, J. R., Jolly, C. A., Krichene, S., Cassoux, P., Faulmann, C., manuscript in preparation.

[127] Rivera, N. M., Engler, E. M., Schumaker, R. R., *J. Chem. Soc. Chem. Commun.* 1979, 184.

[128] Engler, E. M., Martinez-Rivera, N., Schumaker, R. R., *Org. Coat. Plast. Chem.* 1979, *41*, 52.

[129] Engler, E. M., Nichols, K. H., Patel, V. V., Rivera, N. M., Schumaker, R. R., U.S. Pat. 4,111,857 (1978).

[130] Dirk, C. W., Bousseau, M., Barrett, P. H., Moraes, F., Wudl, F., Heeger, A. J., *Macromolecules* 1986, *19*, 266.

[131] Götzfried, F., Beck, W., Lerf, A., Sebald, A., *Angew. Chem. Int. Ed. Engl.* 1979, *18*, 463.

[132] Bohm, M. C., *Phys. Stat. Sol.* 1984, *121*, 255.

[133] Bohm, M. C., *Physica* 1984, *124B*, 327.

[134] Reynolds, J. R., Wang, F., *Macromolecules,* in press.

[135] Mammone, R. J., MacDiarmid, A. G., *J. Chem. Soc. Faraday Trans. 1.* 1985, *81*, 105.

Chapter 2

Ionically Conductive Polymers

Anthony J. Polak

In 1834 Michael Faraday reported that when lead fluoride (PbF_2) was heated red hot, it conducted an electric current and so did the metallic vessel it was heated in. This was a startling observation, since most simple salts are electronic insulators. The high conductivity that Faraday observed is now known to be due to ionic conductivity, and not electronic conductivity. At elevated temperatures (500–700°C) the fluoride anion possesses high ionic conductivity and can easily be transported through the lead fluoride lattice. This was the first report of a high ionic conductivity solid electrolyte. Until the mid-1970s all research on solid ionic conductors had centered on inorganic compounds (primarily ceramics, such as the various phases of alumina, stabilized zirconia, etc.). With the discovery of new ionic conducting polymers by Wright,[21,22] Armand,[2,3] and others, and the numerous advantages that ionic conducting polymers have over ceramics in device fabrication and operating temperature range, it is not surprising that fast ionic conduction in polymers is currently an area of great interest. This interest is a result of a desire both to understand the ionic conduction mechanism in polymers and to use these polymers in applications such as high-energy-density batteries, electrochronic displays, specific-ion sensors, and other electrochemical devices that capitalize on the unique electronic, ionic, and mechanical properties of ionic conducting polymers.

The ability of a polymer or polymer blend to conduct an ion or ions may be either desirable or disastrous. For instance, high ionic conductivity is essential in a polymer that will be used as the solid electrolyte in a polymer battery. Although a less conductive polymer may be acceptable as the solid electrolyte in a galvanic cell sensor, any ionic conductivity (especially from ionic sodium) will be devastating if the polymer is to be used as a dielectric layer in a semiconductor device.[1]

All polymers, and polymer blends, are to some extent ionic conductors. The most common application of an ionic conducting polymer is as the polyelectrolyte in an ion-exchange membrane. However, these resins show extremely low ionic conductivity, typically 10^{-14} (ohm-cm)$^{-1}$ in the absence of water (as a result of extensive ion pairing), and because of their low conductivity are not discussed in this review. Instead, we concentrate only on those polymers and polymer systems in which the ionic conductivity is at least 10^{-6} (ohm-cm)$^{-1}$.

The potential advantages of using ionic conducting polymers as solid electrolytes in batteries were first discussed by Armand.[2,3] The primary advantage is that a viscoelastic electrolyte (such as an ionic conducting polymer) is more flexible than a solid ceramic electrolyte. This flexibility results in a system capable of withstanding large volume changes that occur during repeated charging and discharging cycles while still maintaining intimate contact with the electrodes. In addition, these materials are light, typically with a density near unity, and can easily be processed into thin films, further reducing their resistance and making it practical to use solid electrolytes that have lower conductivities.

Polymers having high ionic conductivities are sometimes called solid electrolytes, fast ion conductors, or superionic conductors (a term disliked by some because it incorrectly implies a conduction mechanism akin to superconductivity, and defended by others as meaning supermobile).[4] The terms *ionic conducting polymer, polymer complex,* and *solid electrolytes* are used interchangeably in this chapter when referring to polymer systems or blends that have a high ionic conductivity. In general, these polymers have two characteristics in common. The first is that the homopolymer or homopolymers themselves do not have significant ionic (or electronic) conductivities and must be doped or blended with other compounds. The second is that the conductivities of these blended or complexed polymers are very high, often approaching those of liquid electrolytes.

If ionic conductivity in a polymer were governed by a simple jump diffusion mechanism, it could easily be shown that the ionic conductivity should not be greater than 10^{-14} (ohm-cm)$^{-1}$, using the approximations associated with Stokes' law and the Nernst–Einstein relation. This value is obtained by combining the Stokes–Einstein equation:

$$D = \frac{kI}{6\pi r\eta} \tag{1}$$

(where D is the diffusion constant, r is the ionic radius, k is Boltzmann's constant, T is temperature, and η is viscosity) with the Nernst–Einstein equation:

$$\sigma = \frac{z^2F^2c(D_+ + D_-)}{RT} \tag{2}$$

(where c is the concentration of mobile ions, z is the charge of the ion, F is Faraday's constant, and D^+ and D^- are the diffusion coefficients of the two mobile species that are present, an anion and a cation). For a situation in which only one mobile species is present, the combination of Equations 1 and 2 predicts that the conductivity is inversely proportional to viscosity[5,6] and is given by

$$\sigma = \frac{z^2F^2c}{N_A 6\pi r\eta} \tag{3}$$

where N_A is Avogadro's number.

Using typical values for the radius of ionic sodium and the viscosity of a polymer (at its T_g), the ionic conductivity of a polymer can be shown to be in the range of 10^{-14} (ohm-cm)$^{-1}$. The conductivity predicted by Equation 3 is clearly not what is observed, as can be seen in Figure 2.1, which shows the range of conductivities achieved by various ionic conducting polymers. Figure 2.1 also shows the relative magnitude of ionic conduction of various ion-conducting polymers compared to ceramic ionic conductors (such as calcia-stabilized zirconia), the liquid electrolytes, and the electronic conduction of metals. The reasons for the significant increase (by about a factor of 10^{10}) in the ionic conductivity of the solid ionic conducting polymer electrolytes will be reviewed.

Ionic conducting polymers can be classified into two groups. The distinction is based on the temperature at which the polymer shows appreciable conductivity relative to its glass transition temperature, T_g. This distinction is found to be directly related to the ionic transport mechanism.[5] The first group is composed of polymers that show appreciable ionic conduction only at temperatures above their glass transition. This group includes all the ionic conducting polymers based on complexes of polyethers and alkali metal salts, such as the classic ionic conducting polymers developed by Wright et al.[7] and Armand et al.[2,3] Also included in this group are the polyphosphazene backbone polymers to which ion-solvating groups have been grafted, such as the polymer MEEP, which is produced by the reaction of sodium 2-(2-methoxyethoxy)ethanol with poly(dichlorophosphazene)[8,15] complexed

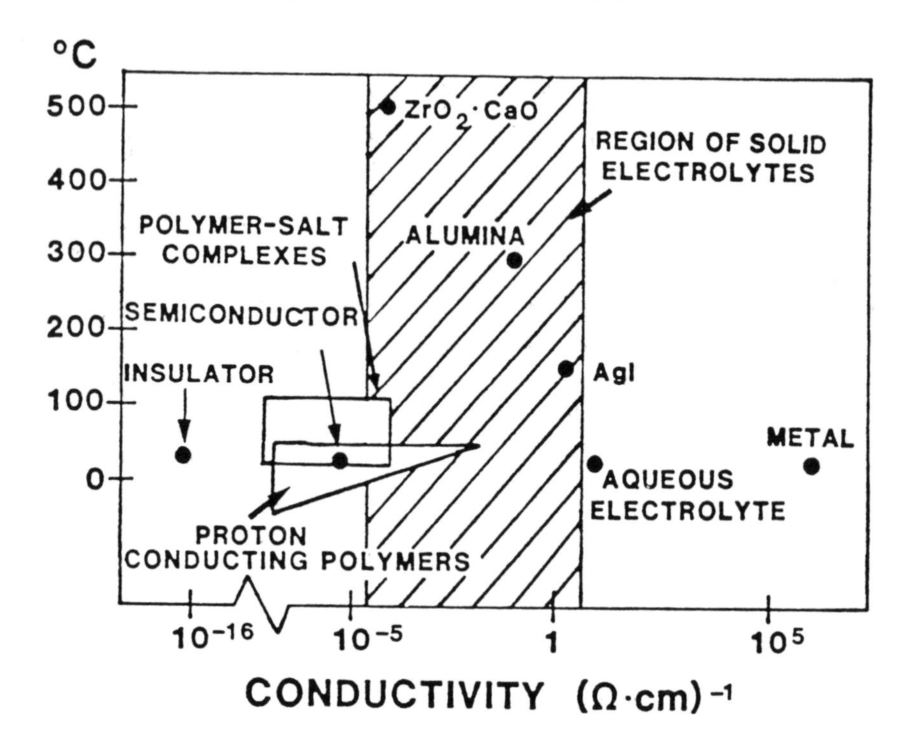

Figure 2.1 Conductivity (both ionic and electronic) of solids at typical temperatures of use.

$$[-N = P(Cl_2)-]_n + 2nNaOC_2H_4OC_2H_4OCH_3 \xrightarrow[THF,\ \Delta]{n-Bu_4\,NBr}$$

$$[-N = P(OC_22H_22OC_22H_44OCH_33)_22-]_n + 2nNaCl \qquad (4)$$

with $AgSO_3CF_3$, and analogues of PEO complexes, such as poly(ethyleneimine) complexed with alkali metal salts. In this group of polymers the ionic conduction process involves a cooperative interaction between the mobile ionic species and the polymer matrix.

The second group consists of polymers that have appreciable ionic conductivity at temperatures (typically, room temperature) below their glass transition temperatures. Nafion[88,89] and blends of poly(vinyl alcohol) and H_3PO_4 (PVA/H_3PO_4)[9] are examples of polymers in this group. In these systems the conduction mechanism can be described by a precolation model in which a highly conductive phase is embedded in a poorer conducting phase (or insulator). Significant conduction does not occur until a critical volume of the highly conductive phase is reached, at which point the highly conductive phase forms a continuous network throughout the polymer matrix.

Theory

Conductivity

The single most important parameter in characterizing an ionic conducting polymer is the temperature dependence of its ionic conductivity. Ionic conductivity is generally measured by ac admittance (or ac impedance, which is the inverse of admittance) techniques. This measurement is based on the measurement of cell admittance (or impedance) taken over a range of frequencies and then analyzed in the complex admittance plane.[17] From the shape of the admittance plot, an equivalent circuit for the polymer system can be determined, along with estimated values of the circuit parameters. The most commonly used equivalent circuit is shown in Figure 2.2a. It consists of a resistor R, which corresponds to the bulk polymer electrolyte resistance, C_g, which is the geometric capacitance, and C_{dl}, which is the double-layer capacitance. The corresponding impedance plot is shown in Figure 2.2b. The nonzero intercept of the semicircular arc with the real axis corresponds to R. As the temperature is increased, this intercept moves to progressively smaller values of resistance. Knowing the temperature dependence of the bulk resistance $R(T)$, the polymer thickness l, and the electrode area A, the temperature dependence of the ionic conductivity can be obtained through the relationship:

$$\sigma = \frac{l}{A\,R(T)} \tag{5}$$

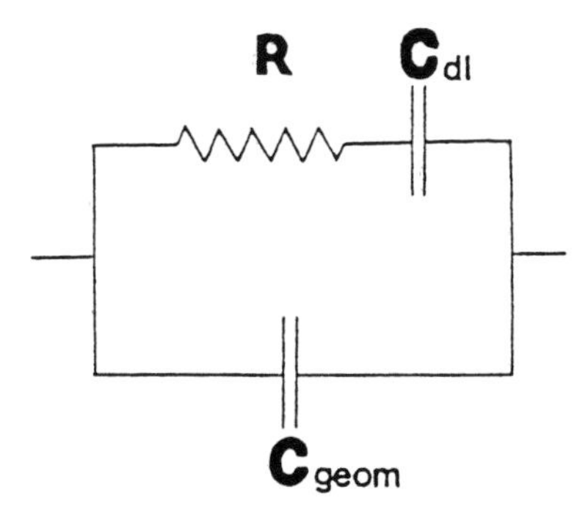

Figure 2.2a Equivalent circuit used to model admittance measurements.

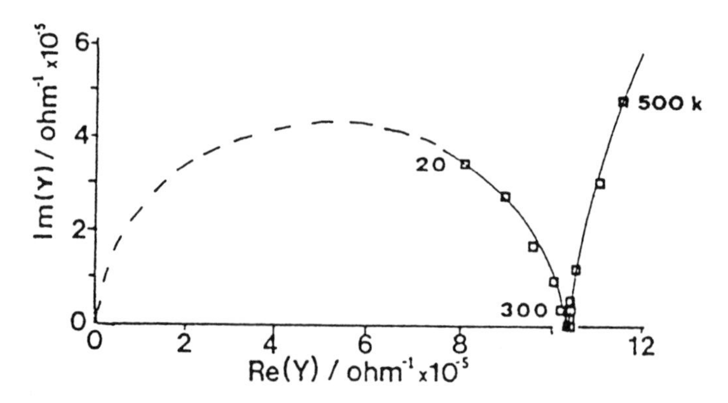

Figure 2.2b Complex admittance plot for PEO $NH_4SO_3CF_3$ 5:1.

The temperature dependence of a physical property X is normally described in terms of an Arrhenius relationship:

$$X = X_0 0 \exp\left(\frac{-E}{Rt}\right) \tag{6}$$

where X_0 is the pre-exponential factor and E is the activation energy. If this relationship holds, then the Arrhenius plot of $\ln(X)$ against $1/T$ will result in a straight line of slope $-E/R$ and a y-intercept of X_0. Many physical processes give linear Arrhenius plots and can be explained in terms of a model in which transitions from an initial energy state to a final state are governed by both temperature-independent parameters (which are contained in the X_0 term, despite the fact that the X_0 term typically has a weak temperature dependence) and the temperature-dependent Boltzmann factor, which describes the probability of surmounting an energy barrier of height E. In normal usage E is an enthalpy or an internal energy rather than a free energy, which results in the entropy term, $\exp(S/R)$, being contained in the preexponential along with the frequency of jump attempts ω and the jump length λ. Using the normal expression for the relationship between the diffusion coefficient D and the activation energy:

$$D = D_0 0 \exp\left(\frac{-\Delta H}{Rt}\right) \tag{7}$$

where

$$D_0 = \left(\frac{\lambda^2 \omega n}{N}\right) \exp\left(\frac{S}{R}\right) \tag{8}$$

n is the number of mobile species, and N is the number of sites. Combining this expression with the Nernst–Einstein equation (Equation 2) results in the expression:

$$\sigma T = \sigma_0 0 \exp\left(\frac{-\Delta H}{Rt}\right) \tag{9}$$

where $\sigma_0 = (\lambda^2 n^2 \omega F^2/NR) \exp(S/R)$. The activation energy can then be obtained from a plot of $\ln(\sigma T)$ against $1/T$. Since $\ln(\sigma T)$ varies slowly with temperature over a small temperature range, a simpler expression is sometimes used:

$$\ln(\sigma T) = \ln(\sigma_0 0) - \ln(T) - \frac{\Delta H}{RT} \simeq \ln(\sigma_0 0) - \frac{\Delta H}{RT} \tag{10}$$

The linear relationship between $\ln(\sigma T)$ and $1/T$ is only observed in those polymer systems in which the polymer has appreciable ionic conductivity below its glass transition temperature, such as in the PVA/ H_3PO_4[9] polymer blend system and in Nafion.[19]

Shown in Figure 2.3 is a typical admittance plot for the PVA/H_3PO_4 system. From the shape of the admittance plot, it is obvious that the simple circuit shown in Figure 2.2a cannot adequately describe this system. The problems associated with applying the equivalent circuit in Figure 2.2a to this admittance plot are

1. The semicircular curve does not have its center on the real axis.
2. The high-frequency spur is not normal to the real axis.
3. The high-frequency intercept does not cross the origin.

Thus the circuit in Figure 2.2a does not adequately model the processes occurring in the polymer and at the polymer–electrode interface. To model this polymer system more accurately, the more complicated equivalent circuit shown in Figure 2.4 was proposed.[57] In this model the polymer consists of two phases. One phase is a "poor" ionic conductor (with high resistance) and the other is a "good" ionic conductor (with low resistance). Also included in this model are the electrode resistances, a resistance associated with moving a charged species from the electrode into the bulk polymer, and a Warburg (Z_w) impedance (generally associated with diffusional processes).

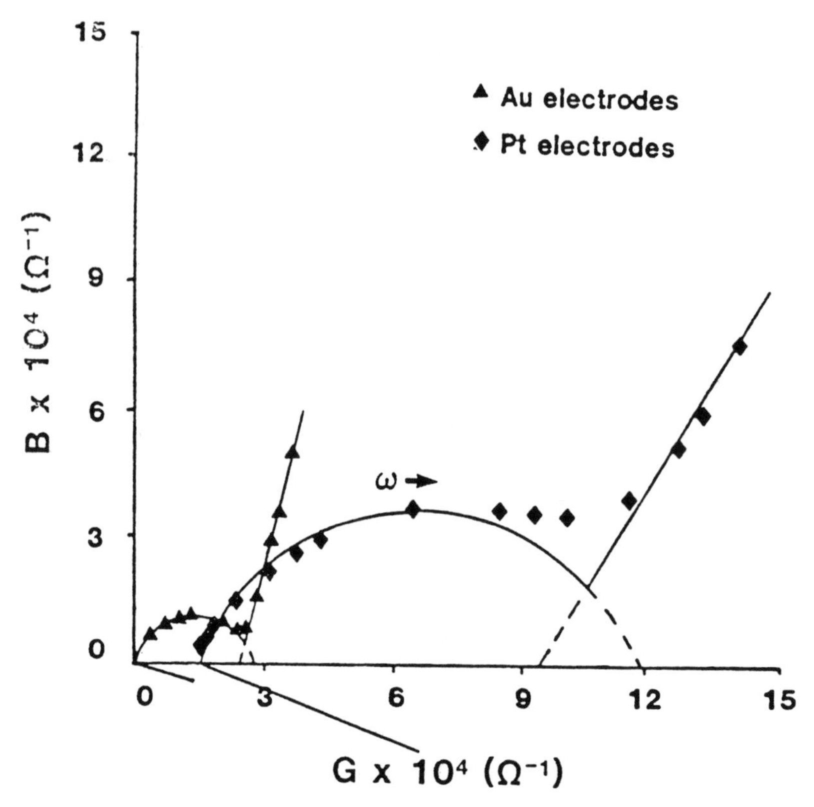

Figure 2.3 Complex admittance plot of PVA/H_3PO_4 (3.8:1), with blocking electrodes (gold) and ohmic electrodes (platinum).

Based on the equivalent circuit in Figure 2.4, the ionic conductivity of a blend of PVA/H_3PO_4 is found to follow an Arrhenius temperature behavior, as shown in Figure 2.5. The temperature dependence is found to be composed of two linear regions, the high-temperature region having an activation energy of approximately 22 kcal/mol, and the low-temperature region having an activation energy of 3.6 kcal/mol. The cause of the break in the Arrhenius plot at −40°C is not known, but may be due to a change in either the conduction mechanism or ionic transport number.

Nafion, swollen with electrolyte solution, also displays a linear Arrhenius behavior. The admittance plots for Nafion, shown in Figure 2.6, were analyzed based on the circuit shown in Figure 2.2a, with the resulting temperature dependence of the ionic conductivity shown in Figure 2.7.

The linear dependence of $\ln(\sigma T)$ versus $1/T$ appears to be the exception rather than the rule when dealing with ionic conducting polymers. For

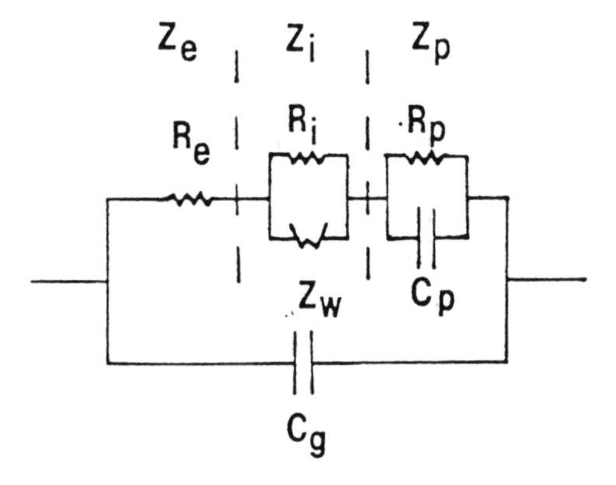

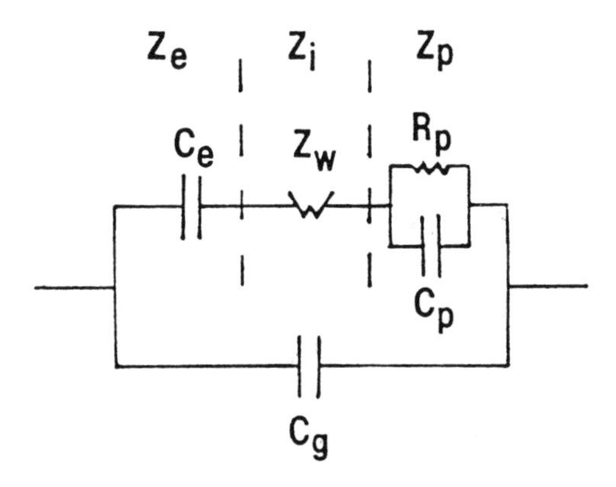

Figure 2.4 Equivalent circuit models for PVA/H_3PO_4 with ohmic electrodes (upper) and blocking electrodes (lower).

amorphous ionic conducting polymers (and polymers in which the ionic conduction occurs through the amorphous phase), the $\ln(\sigma T)$ versus $1/T$ plot is nonlinear (and is typically concave downward),[18,20–29] examples of which are shown in Figures 2.8[30,2] and 2.9a.[2]

Williams et al.[31] showed that in an amorphous polymer above its glass transition temperature a single empirical expression can describe the temperature dependence of all mechanical and electronic relaxation processes. The Williams–Landel–Ferry (WLF) equation relates an electronic or mechanical property of a polymer at a temperature T to the property at a reference temperature T_s through the log of the shift factor and is given by

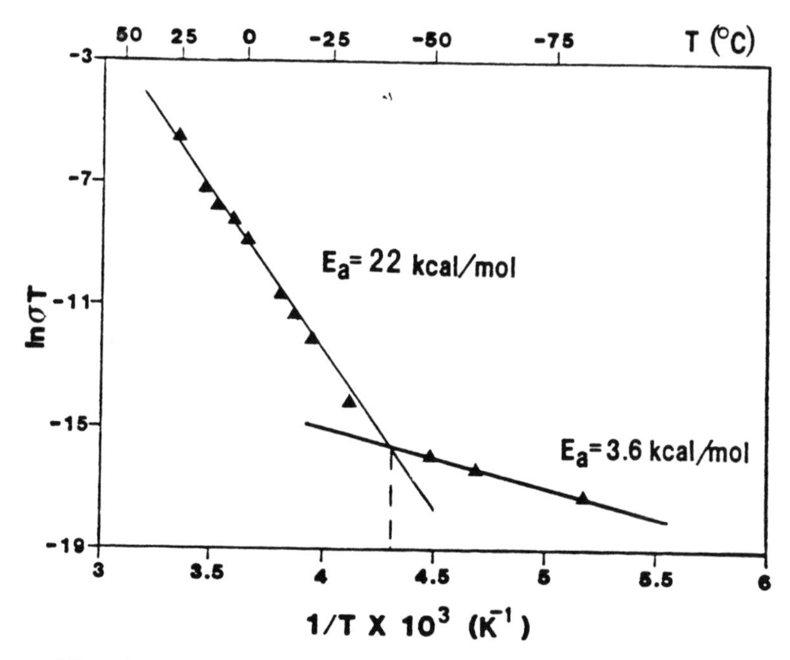

Figure 2.5 Arrhenius plot of ionic conductivity for PVA/H_3PO_4 (3.8:1).

$$\log(a_T) = \frac{-C_1(T-T_s)}{(C_2 + T-T_2)} \tag{11}$$

where a_T is the shift factor, originally defined as

$$a_T = \frac{\eta T_s \rho_s}{\eta_s T \rho} \tag{12}$$

where η and ρ are the viscosity and density at T and η_s and ρ_s are the corresponding quantities at the reference temperature T_s. T_s was originally an empirical parameter, but later work[32] showed that T_s is best interpreted as the thermodynamically limited glass transition temperature. C_1 and C_2 are constants that are found to depend on the reference temperature and the polymer system. It was originally believed that C_1 and C_2 were universal constants, when T_s was taken as the glass transition temperature, with values of 14.44 and 51.6, respectively. From Doolittle's free volume theory of viscosity,[34] the constants C_1 and C_2 can be expressed in terms of the free volume fraction of the glass, f_g, and the free volume expansion coefficient, Δa (assumed to increase linearly with temperature), which is taken to be the difference in the thermal expansion coeffi-

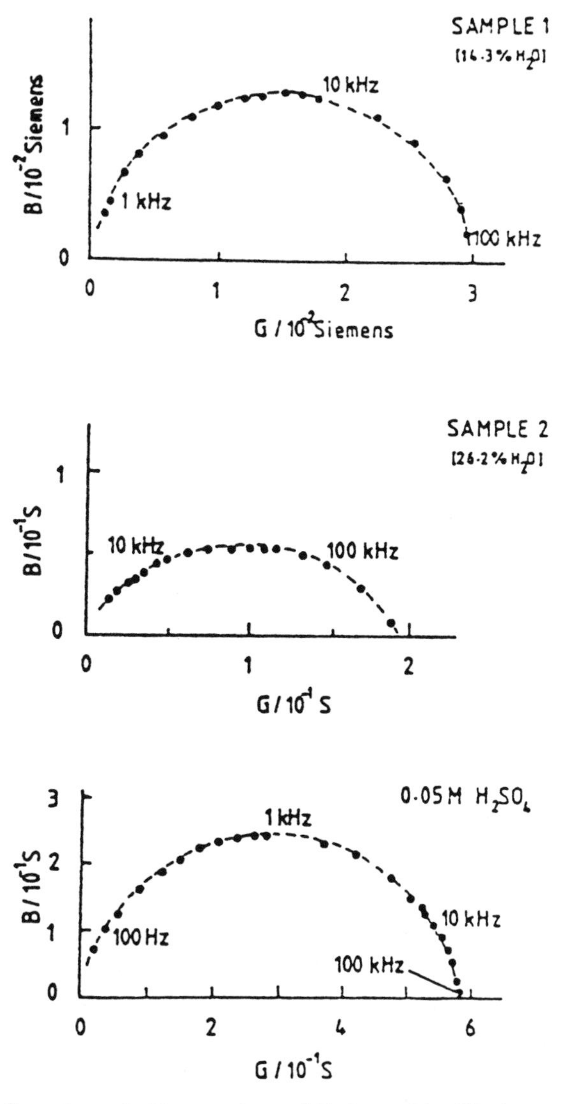

Figure 2.6 Complex admittance plots of Nafion with differing water content (samples 1 and 2) and 0.05M H_2SO_4 (aq) in the test cell.

cient above and below the glass transition temperature. C_1 and C_2 are then given by

$$C_1 = \frac{1}{2.3f_g} \tag{13a}$$

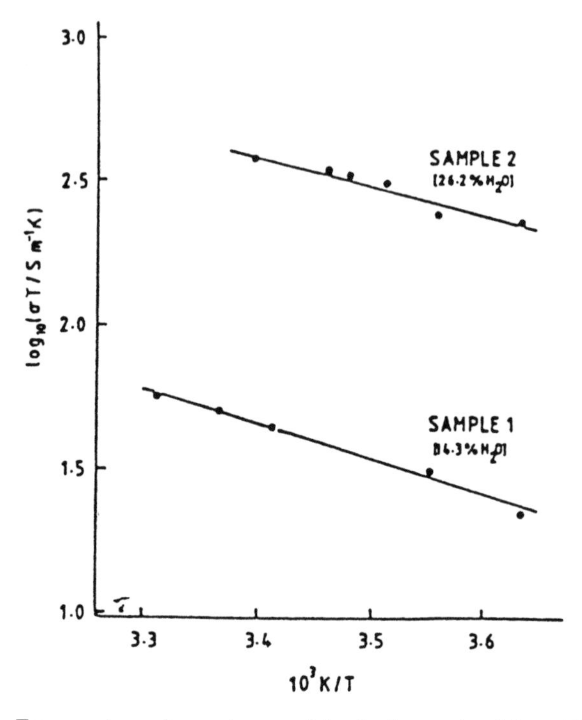

Figure 2.7 Temperature dependence of the ionic conductivity of Nafion.

and

$$C_2 = \frac{f_g}{\Delta a} \tag{13b}$$

If f_g is constant[35] and Δa is a constant (which it is for a large number of polymers), then C_1 and C_2 are indeed universal constants. This, however, appears not to be the case,[36] since C_1 and C_2 are found to depend on the polymer and the mobile ionic species.

Cohen and Turnbull[37] proposed that a critical free volume V^* was required for a particle to diffuse, which results in the WLF constants given by[30,38]

$$C_1 = \frac{\gamma f^*}{2.3 f_g} \tag{14a}$$

and

$$C_2 = \frac{f_g}{\Delta a} \tag{14b}$$

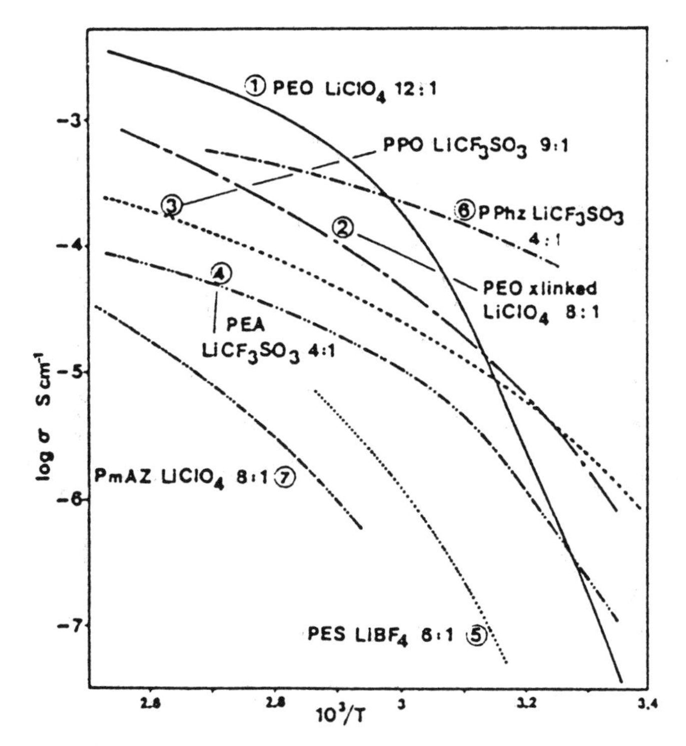

Figure 2.8a Arrhenius plot of the conductivity of various polymer-cell adducts. (1) PEO; (2) cross-linked PEO; (3) poly(propylene oxide); (4) poly(ethylene adipate); (5) poly(ethylene succinate); (6) polyphosphazene; (7) poly(N-methylaziridine).

where γ is a factor that accounts for the overlap of free volume elements and f^* is the critical free volume fraction. Since the critical free volume depends on the diffusing species, this theory helps to explain the materials dependence of the WLF constants. In the case of ionic conducting polymers, the critical free volume is interpreted as the volume required for ionic transport in an amorphous material.

An alternative expression that is often used to describe the ionic conductivity in an amorphous polymer system is based on free volume theory and is given by the Vogel–Tamman–Fulcher (VTF) equation:[33,39]

$$C(T) = A \exp\left(\frac{B}{T - T_0}\right) \tag{15}$$

where $C(T)$ can be any reduced-transport parameter. If $C(T)$ is $DT^{-1/2}$, or $\sigma T^{1/2}$, the following expression for ionic conductivity results:

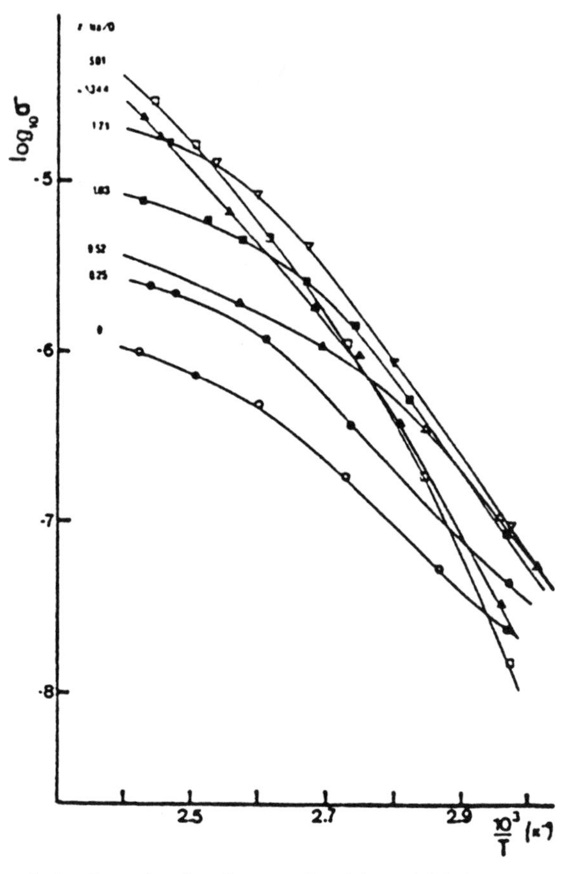

Figure 2.8b Arrhenius plot for the conductivity of PEO 400-based networks containing different amounts of $NaB(C_6H_5)_4$.

$$\sigma = AT^{-1/2} \exp\left(\frac{B}{T - T_0}\right) \tag{16}$$

The $1/(T - T_0)$ is the same temperature dependence developed by Adam and Gibbs[40] based on Gibbs and DiMarzio's[41] configurational entropy model. In this model the polymer is composed of an ensemble of small groups that can undergo cooperative rearrangement at temperatures greater than T_2 ($T_g > T_2$). At T_2 a second-order phase transition would occur if the rate of molecular rearrangement did not become infinitesimal, and the configurational entropy goes to zero. T_2 is thus the temperature at which the cooperatively rearranging region is comprised of the whole sample (or macroscopic parts of it), since at this temperature there is only one available configuration for the entire system (the configurational

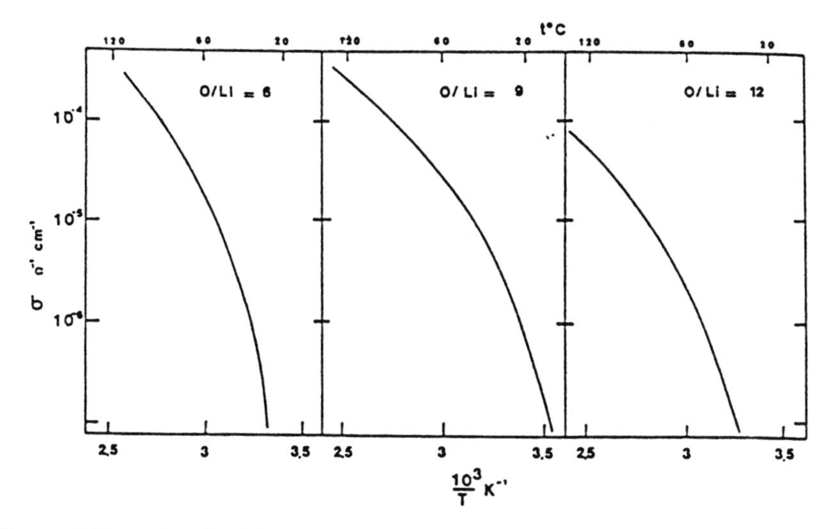

Figure 2.9a Conductivity of the PEO-$LiCF_3SO_3$ complexes.

entropy is zero). This results in an average transition probability $W(T)$, which is given by

$$W(T) = A \exp\left(\frac{-z^* \Delta\mu}{kT}\right) \tag{17}$$

where z^* is the smallest segment that permits a transition, A is a frequency factor, and $\Delta\mu$ is the potential energy barrier hindering the cooperative rearrangement per monomer segment.

The shift factor, from Equation 11, can then be used to express the ratio of transition probabilities at the temperatures T and T_s:

$$\log(a_T) = \log\left[\frac{-W(T)_s}{W(T)}\right] = \frac{-C_{-1}(T - T_{-s})}{C_2 + (T - T_s)} \tag{18}$$

which now results in the WLF constants being given by

$$C_1 = \frac{2.303\, s_c^* \,\Delta\mu}{kT_x \,\Delta C_{-p} \ln(T_s/T_2)} \tag{19a}$$

and

$$C_2 = \frac{T_s \ln(T_s/T_{-2})}{1 + \ln(T_s/T_2)} \tag{19b}$$

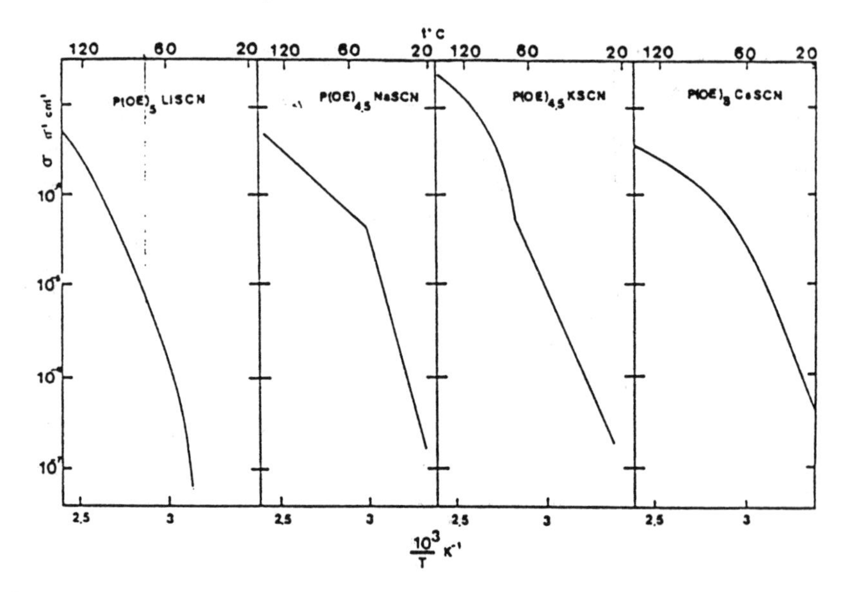

Figure 2.9b Arrhenius plot of the conductivity of the PEO-metal SCN complexes.

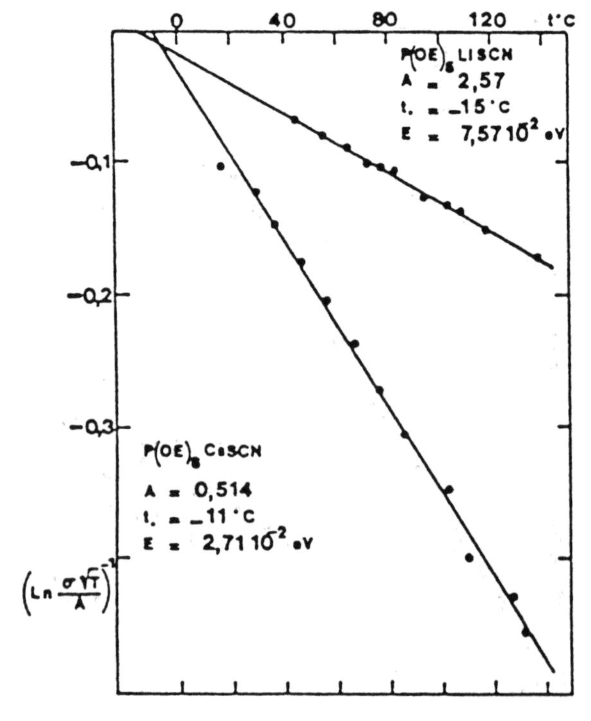

Figure 2.9c VTF plot of conductivity for PEO-LiSCN and PEO-CsSCN.

where s_c^* is the critical configurational entropy and $\Delta\mu$ is the potential energy barrier hindering the cooperative rearrangement per monomer segment.

The power of the WLF equation is that the transport properties of a polymer at any arbitrary temperature T can be correlated to the same physical property at another temperature, typically the glass transition temperature, obtained by viscoelastic measurements. The WLF and VTF equations can be shown to be equivalent, with $C_1C_2 = B$ and $C_2 = T_g - T_0$. When the temperature dependence of the ionic conductivity of a polymer that displays non-Arrhenius behavior is plotted as $\ln(\sigma T^{1/2})$ versus $1/T$, a linear relationship is often observed, as shown in the Arrhenius and VTF (WLF) plots in Figures 2.9b and 2.9c, respectively.

An alternative explanation for the non-Arrhenius behavior of the ionic conductivity plots has also been suggested by several authors.[42,43] Repeating the earlier work performed on PEO–LiSCN, PEO–$LiCF_3SO_3$ and PEO–$LiClO_4$, they found that an Arrhenius plot of the ionic conductivity could be broken up into two linear regions with different slopes (activation energies), as shown in Figures 2.10 and 2.11. The temperature at which the break in the slope occurs corresponds to the melting tempera-

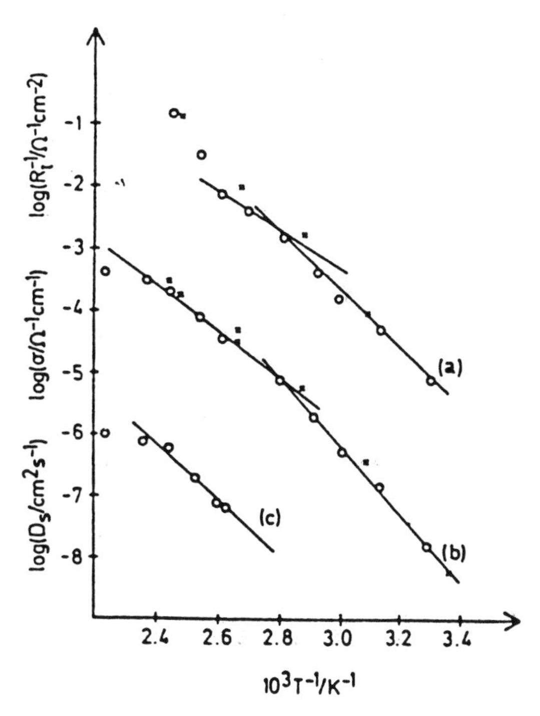

Figure 2.10 Plots of: (a) log(1/R) vs. 1/T; (b) log(σ) vs. 1/T; (c) log(D_s) vs. 1/T.

ture of pure PEO. Based on DSC and NMR results, it is known that the PEO–alkali metal salt complex system is composed of several phases, a pure PEO phase with a melting point of about 70°C and a PEO–alkali salt phase. The apparent contradiction in the interpretation of the ln(σ) versus 1/T plots suggests that above the melting point of the pure PEO phase, a traditional diffusion mechanism is involved in ionic transport. However, the logarithm of the preexponential factor was found to increase approximately linearly with the salt concentration in the polymer, which is not in agreement with classical ionic conductivity models based on a single ion hopping mechanism.

Table 2.1 lists some of the more common ionic conducting polymers, their glass transition temperatures, and their conductivity at specific temperatures. As can be seen from this table, the ionic conducting polymers appear to divide themselves naturally into two groups. The first group is composed of polymers with glass transition temperatures well below the temperature where they have maximum conductivity, such as the complexes of PEO and PPO. The second group is composed of polymers with T_g's above their operating temperatures. At first inspection, it would appear that poly(vinyl pyrrolidone), complexed with $LiSO_3CF_3$, having a T_g of 180°C and a maximum conductivity that occurs at 100°C, would fall in the second group. However, the high ionic conduc-

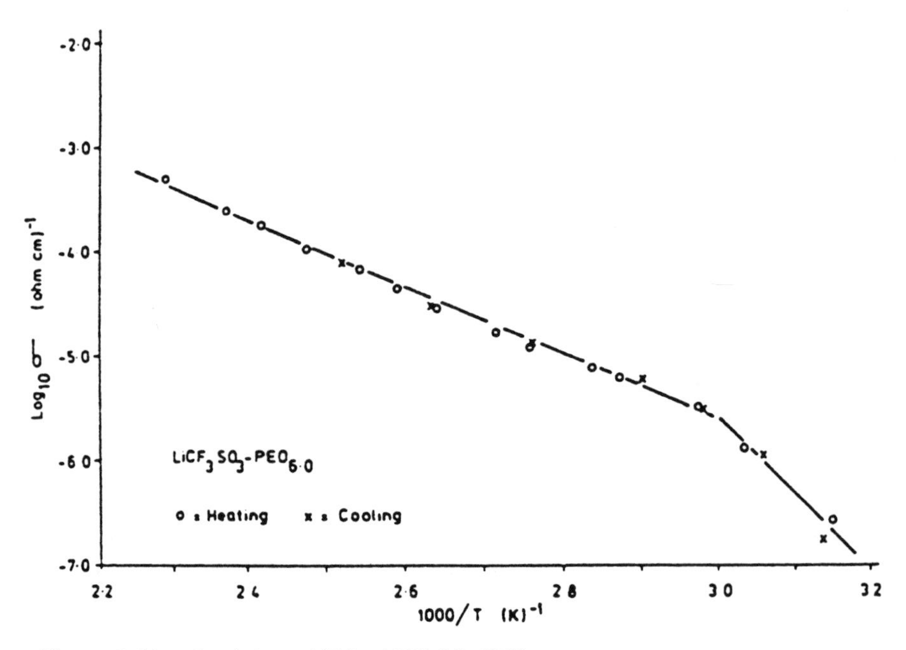

Figure 2.11 Log(σ) vs. 1/T for $LiCF_3SO_3$-$PEO_{6.0}$.

Table 2.1 Glass Transition Temperature and Conductivity of Some Common Polymers

Polymer complex	Homopolymer T_g (°C)	Conductivity $(\text{ohm-cm})^{-1}$ at Temperature T (°C)	Ref.
PPO–$ZnCl_2$	−64		10
$(PPO)_8$–LiI	−64	10^4 at 85	11
$(PPO)_9$–$LiCF_3SO_3$	−64	10^{-4} at 95	11
$(PPO)_{12}$–$NaCF_3SO_3$	−64	10^{-4} at 95	11
(PPO)triol–sodium tetraphenylboride	−64	10^{-4} at 132	16
PEO–KSCN	−60		
PEO–LiSCN	−60		
$(PEO)_4$–NaI	−60		
$(PEO)_{10}$–NaI	−60	10^{-4} at 50	11
$(PEO)_8$–LiI	−60	10^{-4} at 55	11
$(PEO)_{4.5}$–$LiClO_4$	−60	10^{-4} at 110	11
$(PEO)_8$–$LiClO_4$	−60	10^{-4} at 50	11
$(PEO)_{4.5}$–$LiCF_3SO_3$	−60	10^{-4} at 120	11
$(PEO)_8$–$LiCF_3SO_3$	−60	10^{-4} at 100	11
$(PEO)_{14}$–$LiCF_3SO_3$	−60	10^{-4} at 85	11
$(PEO)_{4.5}KCF_3SO_3$	−60	10^{-4} at 70	11
Polyphosphazene	−83		
(MEEP)–$(AgSO_3CF_3)_{0.25}$		9.7^{-3} at 70	12
(MEEP)–$[Sr(SO_3CF_3)_2]_{0.25}$			12
(MEEP)–$(NaSO_3CF_3)_{0.25}$			12
(MEEP)–$(LiSO_3CF_3)_{0.25}$			15
Poly(pentamethylene sulfide)–$(AgSO_3CF_3)_{.25}$		5×10^{-8} at 45	13
PEI–NaI			
PEI–$LiClO_4$	−35	10^{-4} at 109	18
PEI–$LiBF_4$	−35	10^{-4} at 74	18
Poly(vinyl pyrrolidone)–PEG_2–$LiSO_3CF_3$	180*	5×10^{-5}	13
Poly(ethylene succinate)$_3$–$LiBF_4$–1		3.4×10^{-6} at 65	14
PVA/H_3PO_4	70	10^{-5} at 24	57
Nafion	>150	10^{-3} at 24	61–63

* PEG plasticizes the PVP and reduces the T_g to −55°C.

tivity of this polymer complex is only achieved after it is plasticized with poly(ethylene glycol)-PEG, which lowers the glass transition to approximately −55°C.

Based on Equations 9 and 14, ionic conductivity is expected to increase with temperature for polymer systems that display both Arrhenius and non-Arrhenius behavior (see Figures 2.7 and 2.9). Ionic conductivity differs as a function of acid content (for the PVA/H_3PO_4 system) and salt content (for the polyethers, PEI, MEE, etc.). In PVA/H_3PO_4, conductivity is low and relatively constant at low acid concentrations. At a critical acid concentration an abrupt change in conductivity occurs, as shown in Figure 2.12.[57] Increasing the acid concentration further does not markedly increase the ionic conductivity. This is precisely the behavior that would be expected for a system whose conductivity is based on a percolation mechanism. This will be addressed in a subsequent section.

For polymer–salt complexes, the ionic conductivity as a function of salt content typically goes through a maximum, as shown in Figure 2.13 for $AgCF_3SO_3$ in MEEP and $LiCF_3SO_3$ in PPO. The conductivity maximum can be predicted using Equation 16, where A is proportional to the number of charge carriers and the T_0 term in the exponent is closely related to the glass transition temperature of the polymer salt. The conductivity of the polymer–salt complex increases with salt content because of the increase in charge carrier concentration. However, as the salt concentration increases, so does the glass transition temperature, resulting in restricted segmental motion of the polymer and decrease in the ion mobility.

Mobility

A fundamental understanding of the high ionic conductivity observed in polymer electrolytes requires knowledge of the mobility and transfer

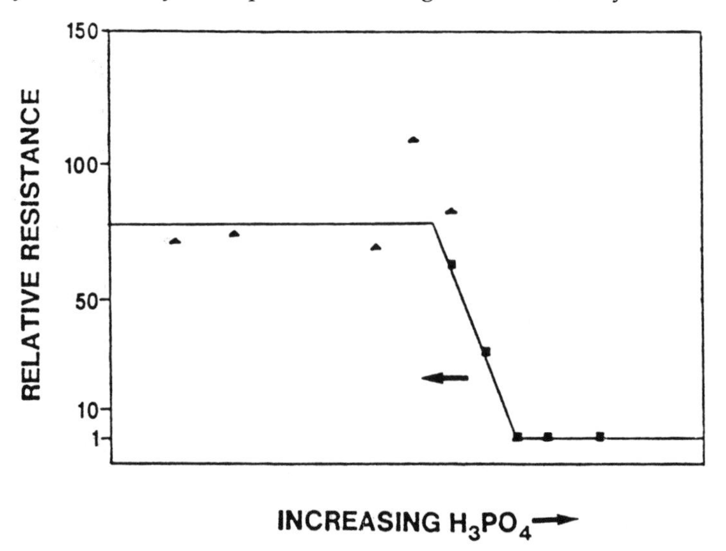

Figure 2.12 Changes in relative resistance and T_g for the proton conducting polymer blend of PVA/H_3PO_4, as a function of acid content.

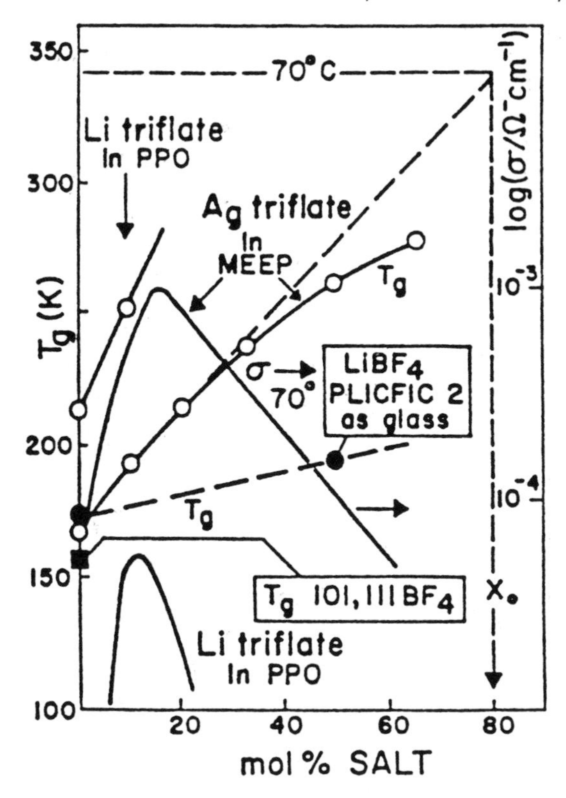

Figure 2.13 Glass transition temperature for polymer salt solutions and PLICFICs, and conductivity versus salt contents.

number of the mobile ion(s). The ionic transfer number t_i is the fraction of the total current carried by the mobile ion. Since most ionic conducting polymer electrolytes are complexes of a host polymer and an alkali metal salt, both the cation and anion can migrate, and both the cationic transfer number t_+ and anionic transfer number t_- must be known. Knowledge of the ionic mobility can be used to determine the ionic transfer number,[44] since the ionic transfer number is related to the ionic mobility through the expression

$$t_+ = \frac{\mu_+}{\mu_+ + \mu_-} \tag{20}$$

where μ_+ and μ_- are, respectively, the cationic and anionic mobilities (assuming that there is no electronic conductivity). The ionic mobility is also related to the ionic conductivity through the expression

$$\sigma = \sum_i \mu_i e n_i z_i \tag{21}$$

where n_i is the number of ionic carriers, z_i is the valency of the ionic carriers, and e is the electric charge. If both monovalent cations and anions contribute to the ionic conductivity, then Equation 21 becomes

$$\sigma = \mu_+ e n_+ + \mu_- e n_- \tag{22}$$

There has been relatively little published data on ionic mobilities and ionic transfer numbers in polymer complexes. Using NMR techniques Armand et al.[2] measured the diffusion coefficient of Li^+ ions in PEO–LiSCN salt complexes and determined that the transfer number was close to unity. However, using low-frequency impedance measurements, Sorensen and Jacobsen[42] reported that the transfer number of Li^+ ions in PEO–LiSCN complexes is about 0.5. PEO is known to exist in two phases, an amorphous phase and a crystalline phase. The high ionic conductivity of the polyether complexes is attributed to ionic motion in the amorphous region. It appears that much of the ambiguity in the conductivity, mobility, and transfer data results from variations in sample preparation. Polymer complexes vary in crystallinity from laboratory to laboratory. More recently, Bouridah et al.,[46] using an electrochemical technique based on concentration cell EMF measurements, determined that at 90°C, PEO–LiI and $PEOLiClO_4$ complexes have a t_{Li^+} of 0.3, while for $PEO–LiCF_3SO_3$ the transfer number is close to 0.7. Using NMR techniques, Bhattacharja, et al.[55] measured the cationic and anionic transfer number in the $PEO–LiCF_3SO_3$ complex and found t_+ ranging from 0.337 at 428 K to 0.404 at 448 K, given by the expression

$$t_{Li^+} = \frac{1}{1 + 3.276 \times 10^{-3} \exp(0.236\ \mathrm{eV}/kT)} \tag{23}$$

If electronic conductivity in this material is negligible, as is generally assumed, then $(1 - t_{Li^+})$ should yield the transport number for the triflate ion ($CF_3SO_3^-$). Both transport numbers for the amorphous phase are plotted as a function of temperature in Figure 2.14.

Also using NMR, Berthier et al.[54] have demonstrated that in the PEO–alkali metal salt complexes, $PEO_8/LiCF_3SO_3$ and PEO_{10}/NaI, ionic motion takes place only in the elastomeric phase. Knowing the number of charge carriers and assuming that only cationic conductivity exists in the polymer complex, Berthier et al. deduced the diffusion coefficient from the conductivity of the amorphous phase.

An alternative method for determining the diffusion coefficient is by

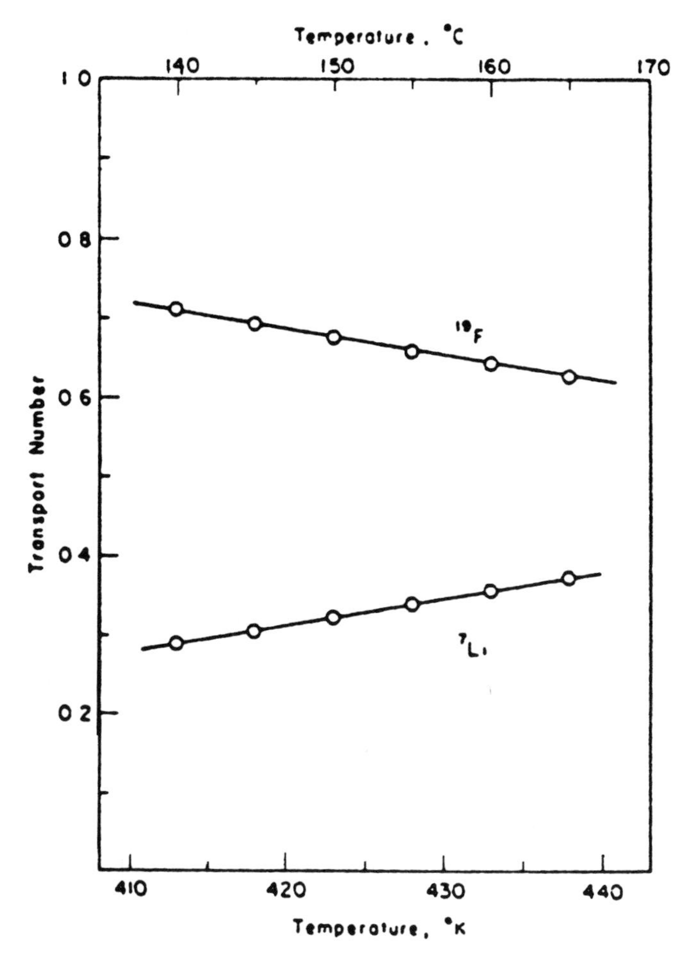

Figure 2.14 Cationic and anionic transport numbers as a function of temperature.

directly measuring the ionic mobility and then relating the ionic mobility to the diffusion coefficient through the Nernst–Einstein equation:

$$\mu = \frac{ZeD}{kT} \tag{24}$$

Examples of the diffusion coefficients for selected polyether–alkali metal salt complexes obtained by NMR, isothermal transient ionic current (ITIC), etc., are listed in Table 2.2.

Watanabe et al.[47–49] proposed an alternative method for evaluating the

Table 2.2 Diffusion Coefficients for Selected Polyether–Alkali Metal Salt Complexes

Polymer System	Temperature (°C)	Ion	D (cm^2 s^{-1})	Ref.
PPO/$LiClO_4$–([$LiClO_4$]/[PO unit] = 0.042)	−1	Li	6.9×10^{-11}	48
PPO/$LiClO_4$–([$LiClO_4$]/[PO unit] = 0.076)	11	Li	5.8×10^{-11}	48
PEO/$LiClO_4$–([$LiClO_4$]/[EO unit] = 0.01)	−43	Li	9.9×10^{-12}	49
PEO/$LiClO_4$–([$LiClO_4$]/[EO unit] = 0.10)	−15	Li	1.9×10^{-13}	49
PEO/LiCF3SO_3–([$LiCF_3SO_3$]/[EO unit] = 0.12)	67	Li	8.4×10^{-9}	54
PEO/$LiCF_3SO_3$–([LiCF3SO_3]/[EO unit] = 0.12)	67	Li	4.7×10^{-9}	55
PEO/$LiCF_3SO_3$–([$LiCF_3SO_3$]/[EO unit] = 0.12)	67	F	4.8×10^{-8}	55
PEO/NaI–([NaI]/[EO unit] = 0.1)	67	Na	3.3×10^{-9}	54
PEO/$LiClO_4$–([$LiClO_4$]/[EO unit] = 0.22)	67	Li	3.6×10^{-10}	56
PEO/$LiClO_4$–([$LiClO_4$]/[EO unit] = 0.12)	67	Li	1.6×10^{-8}	56

ionic mobility in polymer complexes based on the ITIC technique. This technique has been used previously to determine the ionic mobility in dielectrics, such as SiO_2 films in MOS capacitors[50–52] and insulating polymer films.[53] Watanabe used ITIC to determine the ionic mobility (of both cations and anions) in PPO/$LiCLO_4$, poly(ethylene succinate)/LiSCN, and PEO/$LiCLO_4$ complexes. Samples for ITIC were prepared with blocking (platinum) electrodes (platinum forms blocking electrodes for the alkali metal cations). A dc voltage of 2.5–3 V is applied across the polymer for an appropriate period of time (15–120 min) to obtain charge separation at the two electrodes. The polarity of the applied voltage is then quickly reversed, and the current is measured as a function of time. The time dependence of the monitored current is found to have a peak at a time (referred to as the time of flight) τ, which can be related to the ionic mobility by the expression

$$\mu = \frac{d^2}{\tau(V - V_{eff})} \tag{25}$$

where d is the film thickness, τ is the time at which the current peaks in the current versus the time curve, V is the applied voltage, and V_{eff} is the effective voltage across the polymer complex. $V - V_{eff}$ is assumed to be the average voltage across the polymer between times $t = 0$ and $t = \tau$. Shown in Figures 2.15–2.17 are the ITIC curves for poly(ethylene succinate)/LiSCN, PPO/$LiClO_4$, and PEO/$LiCLO_4$, respectively.

Ionic Conductivity Based on a Percolation Mechanism

As was discussed earlier, ionic conducting polymers such as Nafion swollen with an electrolyte solution and PVA/H_3PO_4 blends fall into the

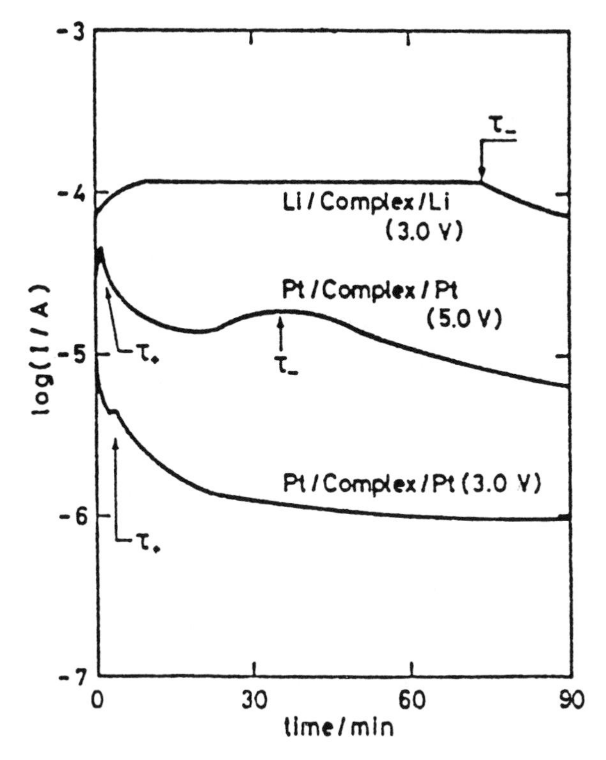

Figure 2.15 ITIC curves for poly(ethylene succinate)-LiSCN complex after application of a dc voltage step for 120 min. and the reversal of the applied polarity.

broad category of polymers that exhibit appreciable conductivities at temperatures below their T_g, while the polyether–alkali metal complexes, polysuccinate–metal complexes, and polyphosphazine–metal complexes exhibit high ionic conductivity at temperatures well above the glass transition of the polymer complex. In addition, the ionic conductivity of Nafion and blends of PVA/H_3PO_4 exhibit a linear Arrhenius behavior, while the temperature dependence of the ionic conductivity of the polyether complexes has a nonlinear Arrhenius behavior.

Experimental data[57,58] and theoretical calculations[59,60] indicate that Nafion and blends of PVA/H_3PO_4 represent a multiphase system, composed of ionic conducting clusters dispersed in a continuous less conductive or insulating polymer matrix. Appreciable ionic conductivity is observed only when a critical volume fraction of the ionic conducting clusters is reached, and the material actually undergoes an insulator to conductor transition. Figure 2.12 shows the resistance of a PVA/H_3PO_4 polymer blend as a function of H_3PO_4. As the volume fraction of H_3PO_4 increases, there is initially very little change in the polymer resistance.

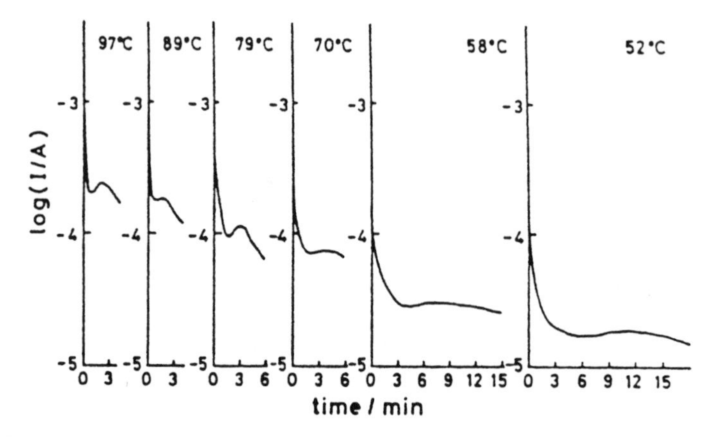

Figure 2.16 Change in current with time after the application of a 2.5 Vdc potential for 15–60 min. in one direction and the reversal of the applied voltage polarity for the PEO-$LiClO_4$ complex ([$LiClO_4$]/[PO unit]=0.076) at various temperatures.

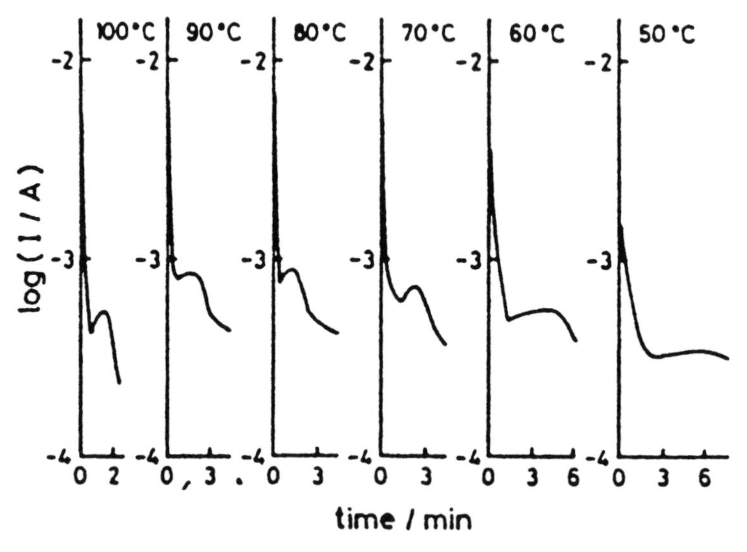

Figure 2.17 Transient ionic current for PEO-$LiClO_4$ complexes of [$LiClO_4$]/[EO unit]=0.01.

At this point the highly conductive clusters are well separated and do not form a continuous path through the polymer. At this stage the resistance of the polymer blend is controlled by the poorer-conducting polymer phase. At a critical volume fraction of H_3PO_4, the highly conductive phase forms a continuous path across the polymer, and a sudden decrease in resistance is observed. This behavior can be explained by using a percolation model for ionic conductivity. A percolation model can also be used to describe the conductivity of Nafion membranes.[61–63]

Recently, Druger et al.[64–66] proposed a model for diffusion (ionic conductivity) in a dynamically disordered polymer system, such as the polyether–alkali metal salt system, based on a dynamic-bond percolation model.

In earlier static percolation models, transport occurs by hops between sites that are either forbidden or allowed with specific fixed probabilities. This static percolation model can be used to describe the conductivity of both Nafion membranes and blends of PVA/H_3PO_4. In these cases ionic conductivity occurs at temperatures below the glass transition of the polymer, and the polymer can be considered a rigid matrix, with hopping sites that either can or cannot accept a migrating ion. Since the polymer is below its glass transition, the density and distribution of hopping sites does not change appreciably with time. In the dynamic-bond percolation model the distribution, density, and availability of a potential hopping site to accept an ion is not fixed, but changes with time as a result of the structural evolution of the host polymer. Systems that appear to be characterized by this kind of dynamic disorder include those displaying electron or polaron motion through a liquid medium and proton movement by means of a Grotthus mechanism.[67–69] In systems that can be modeled by a dynamic-bond percolation model, two characteristic time scales are observed. The first is the time between hops that do not require the rearrangement of the polymer. The second is associated with the time required for the host medium to reorganize (or renew) and thereby provide a new set of preferred pathways for hopping. Druger et al. found that for observation times t much greater than a critical renewal time τ_{ren}, the process is diffusive, independent of the percentage of available bonds and whether or not diffusive behavior is found within a single renewal time ($t < \tau_{ren}$). When the average time between hops, τ_{hop}, is much less than τ_{ren}, the migrating ions (in one dimension) are equally distributed along connected sites, and a fixed mean square distance, which depends only on the fraction of filled bonds, is found. In this case, the problem is reduced to the usual static percolation case.

Applications

The solid-state polymer electrolyte is a new concept in solid state electrochemistry and polymer chemistry. These polymers, which have conductivities greater than 10^{-4} (ohm-cm)$^{-1}$ can routinely be fabricated in the form of thin films with a thickness of less than 1 μm, giving a material with a sheet resistance of less than 1 ohm. This has prompted many investigators to examine the use of these polymers for a host of electrochemical applications, including batteries, fuel cells, sensors, electromagnetic interference shielding, photoelectric cells, electrochromic displays, and ion pumps. The worldwide interest in polymer electrolytes stems from not only their high ionic conductivity, but also their favorable processing parameters, their flexibility and conformability, and their ability to accommodate large-volume changes during the ion–electron exchange process.

Hydrogen Sensor

The use of ionic conducting solid electrolytes as high-temperature fuel cells in galvanic and potentiometric cells to measure kinetic and thermodynamic properties and as monitors is well known.[70–73] Blends of poly(vinyl alcohol) and H_3PO_4 have been shown to be efficient conductors of protons over a wide temperature range (−40°C to 50°C), and as a result can be used as the basis of a hydrogen sensor.[9] The EMF of the potentiometric cell, shown in Figure 2.18, depends on the difference in free energy at each electrode. In addition, the cell voltage depends on the relative conductivities of electrons, holes, and ions across the polymer.

If the polymer is used to separate two regions of differing hydrogen pressure, or concentration, and if each polymer surface has an appropriate electrode, one on which hydrogen can be dissociated and associated (such as platinum or palladium), an EMF will be generated across the polymer. At the anode, molecular hydrogen at a partial pressure P''_{H_2} and chemical potential μ''_{H_2} is dissociated and incorporated into the polymer matrix:

$$H_2(P''_{H_2}) \rightarrow 2e^- + 2H^+ \tag{26}$$

whereas at the cathode the reverse reactions takes place:

$$2H^+ + 2e^- \rightarrow H_2(P'_{H_2}) \tag{27}$$

with two protons and two electrons combining to form molecular hydrogen at a hydrogen partial pressure P'_{H_2}. The net result is the transport of hydrogen from anode to cathode, from a region of high hydrogen concentration to one of lower hydrogen concentration,

$$H_2(P''_{H_2}) \rightarrow H_2(P'_{H_2}) \tag{28}$$

For a reversible process, a statement of the combined first and second laws of thermodynamics is

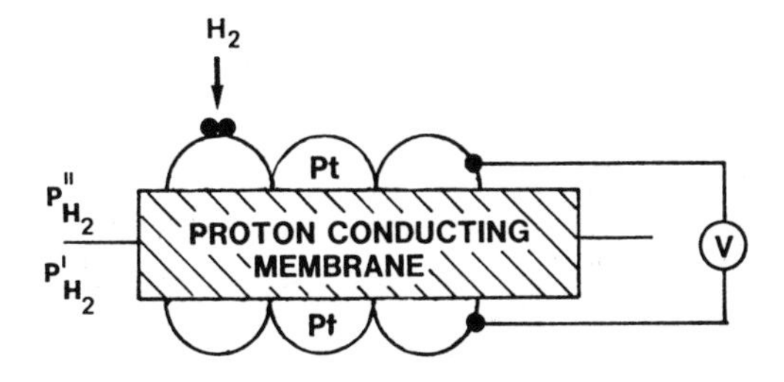

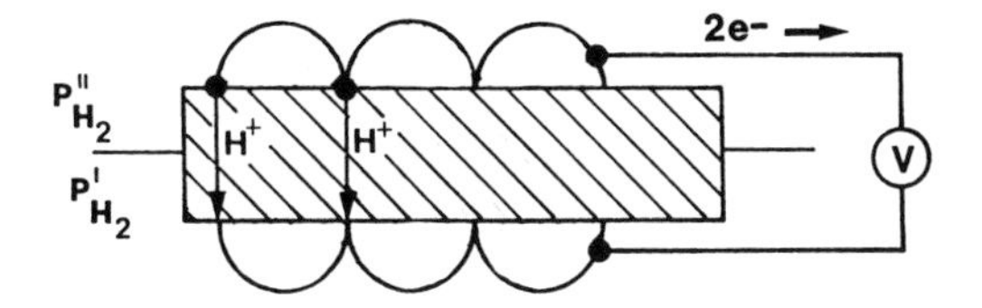

$$\text{EMF} = \frac{RT}{2F} \ln\left(\frac{P_{H2}'}{P_{H2}''}\right)$$

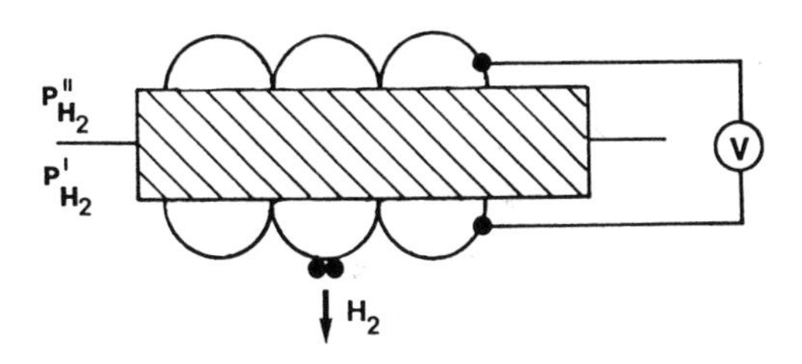

Figure 2.18 Schematic drawing of a proton conducting polymer with two platinum electrodes, separating two regions of differing hydrogen concentration.

$$dF = V\,dP - S\,dT - dW \tag{29}$$

At constant temperature T and total pressure P the free-energy change dF is given by all non–P–V work, that is $-dW$. The free-energy change on passing 1 faraday of electrical charge **F** across the electrolyte is equal to the reversible electrical work, which is given by

$$dF = -\mathbf{F}\,dE \tag{30a}$$

or

$$\Delta F = -F\,\Delta E \tag{30b}$$

The free energy of a reaction is also related to the chemical potentials of the reactants and products by

$$\Delta F = \mu_{H_2'} - \mu_{H_2''} = -2\mathbf{FE} \tag{31}$$

The factor of 2 in the preceding equation comes from the requirement that two electrons be transferred for each hydrogen molecule. The chemical potential can also be expressed in terms of the hydrogen partial pressure:

$$\mu_{H_2} = \mu^{\circ}_{H_2} + RT \ln \frac{P_{H_2}}{P^{\circ}_{H_2}} \tag{32}$$

Combining Equations 31 and 32 results in the Nernst equation:

$$E = \frac{RT}{2F} \ln \frac{P'_{H_2}}{P''_{H_2}} \tag{33}$$

If the electrolyte exhibits mixed conductance (both electronic and mixed ionic conductance), the ionic transfer number t_i must be introduced into Equation 33:

$$E = \frac{RTt_i}{2\mathbf{F}} \ln \frac{P''_{H_2}}{P'_{H_2}} \tag{34}$$

where the ionic transfer number is defined as

$$t_i = \frac{\sigma_i}{\sigma_i + \sigma_e} \tag{35}$$

where σ_i is the ionic conductance and σ_e is the electronic conductance. From Equation 2.34, it can be seen that at constant T (and $t_i = 1$) the EMF developed across the electrolyte depends only upon the hydrogen partial pressure ratio across the electrolyte. A hydrogen sensor can now be devised by keeping the hydrogen partial pressure at one electrode (reference) fixed and measuring the EMF as a function of the unknown hydrogen concentration at the other (working) electrode. Using a reference hydrogen pressure of 1 atm, the hydrogen concentration in a gas stream was measured over nearly four orders of magnitude.[9] Figure 2.19 shows a plot of the measured hydrogen concentration in a hydrogen–nitrogen gas stream versus the actual hydrogen concentration, from a hydrogen concentration of about 200 ppm to 1 atm. In general, the measured hydrogen concentration is within 1% of the actual value. The response of this sensor is on the order of 6 seconds (the time required to attain

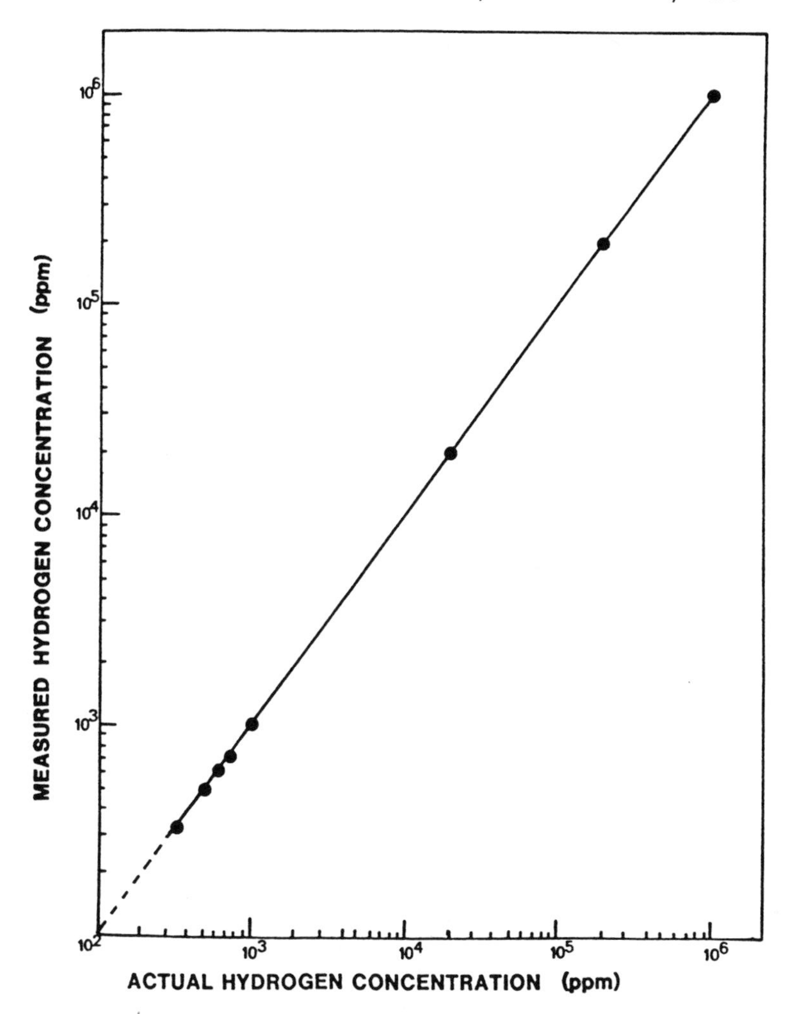

Figure 2.19 Measured hydrogen concentration vs. actual hydrogen concentration in a hydrogen/nitrogen gas stream.

90% of its final value). Figures 2.20a and 2.20b show the response of the hydrogen sensor to a change in hydrogen concentration from 100% to 10% hydrogen. In addition to being very accurate and having high resolution, this sensor also has a high degree of selectivity in that it responds only to hydrogen, or to compounds that generate or consume hydrogen, in the presence of a platinum catalyst. This high degree of selectivity is the result of the selective nature of the proton-conducting polymer, that is, the polymer conducts only protons, and the selectivity introduced into the system by the choice of platinum electrodes. The

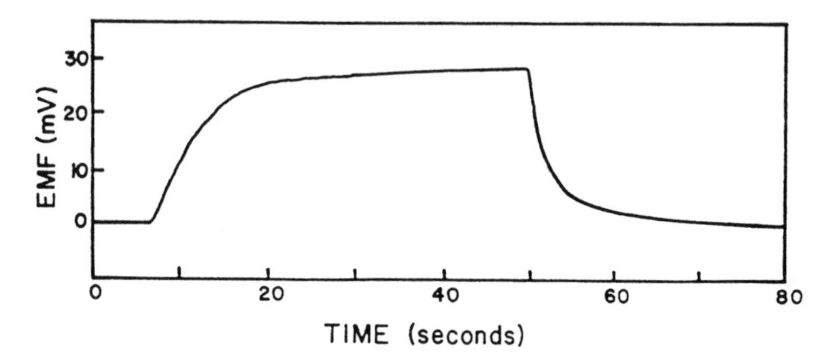

Figure 2.20a Response time measurement of the polymeric hydrogen sensor to changes in hydrogen concentration at the working electrode. The hydrogen concentration was changed from 100% to 10% to 100% hydrogen.

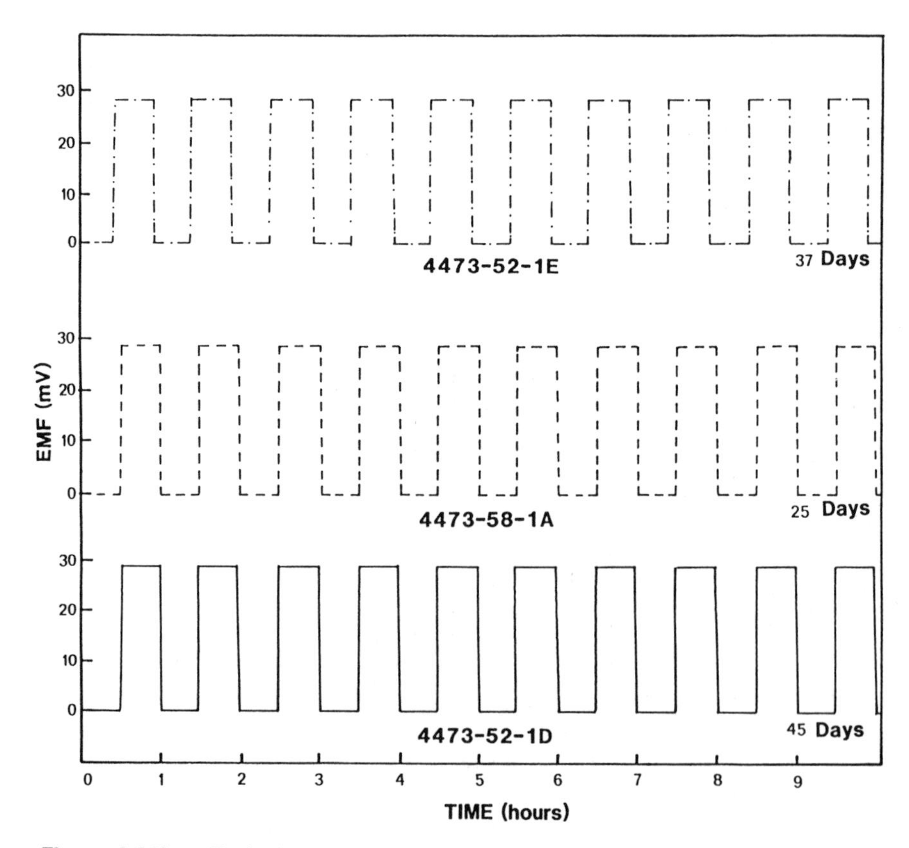

Figure 2.20b Typical response time curves of the polymeric hydrogen sensor after extended times online.

performance (accuracy, response time, resolution, etc.) of this sensor appears to be unaffected by most common gases (methane, ethane, butane, CO_2, N_2, etc.). However, compounds that poison a platinum catalyst or interfere with the dissociation or association of hydrogen on platinum result in the generation of an erroneous voltage. For instance, it is well known that CO and hydrogen compete for the same surface sites on a platinum catalyst, and that CO is preferentially adsorbed. If CO is present in a gas stream at a high enough concentration (>0.1% at room temperature), the platinum surface will be covered with CO, as shown schematically in Figure 2.21, and no hydrogen can then be adsorbed and dissociated or associated. However, once the CO is removed from the gas stream, the adsorbed CO desorbs from the surface, leaving the exposed platinum, which is now available for hydrogen adsorption, dissociation, and association. An on-line hydrogen monitor based on the proton-conducting polymer PVA/H_3PO_4, which operates on the principle described earlier, is shown in Figure 2.22.[45]

An alternative approach to using a gaseous reference source is to use a solid-state reference. Materials that form hydrides, such as palladium, palladium alloys, tungsten trioxide, and so on, have a compositional range over which an increase in hydrogen concentration does not result in an appreciable increase in the hydrogen partial pressure,[74] as is shown in Figure 2.23. By forming a palladium hydride–polymer–platinum electrode system, as shown in Figure 2.24, a solid-state hydrogen sensor has been developed.[9] Initially, several thousand angstroms of palladium were sputter-deposited onto an alumina substrate, and the PVA/H_3PO_4 solution cast onto the surface. After the water had evaporated, the 400-Å platinum electrode was sputter-evaporated onto the polymer surface. The platinum and palladium electrodes were then electrically connected and the device was placed in a closed chamber at a hydrogen pressure of 1 atm. This resulted in the migration of protons from the platinum electrode across the polymer to the palladium electrode, where palladium hydride was formed. After a suitable time (approximately 45 min), the partial pressure of hydrogen at the palladium hydride electrode was that associated with the plateau region of the palladium hydride composition–partial pressure diagram (see Figure 2.23). This corresponded to a hydrogen partial pressure of 1.12×10^{-2} atm, compared to a theoretical value of 1.18×10^{-2} atm. Changing the hydrogen pressure in the test cell resulted in a change in sensor EMF, in agreement with the Nernst equation. The response of the sensor to changes in hydrogen pressure (concentration) at the working electrode is shown in Figure 2.25.

Electrochromic Displays

Electrochromic displays are devices that undergo a reversible color change as a result of an applied electric field or current. An electrochromic

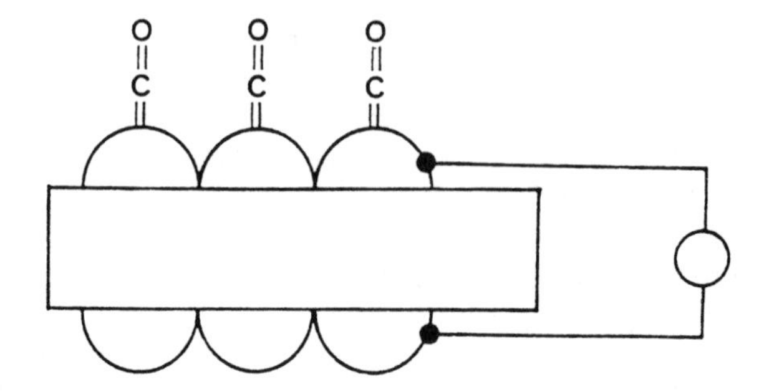

Figure 2.21a Schematic representation of the platinum working electrode chemisorbing CO.

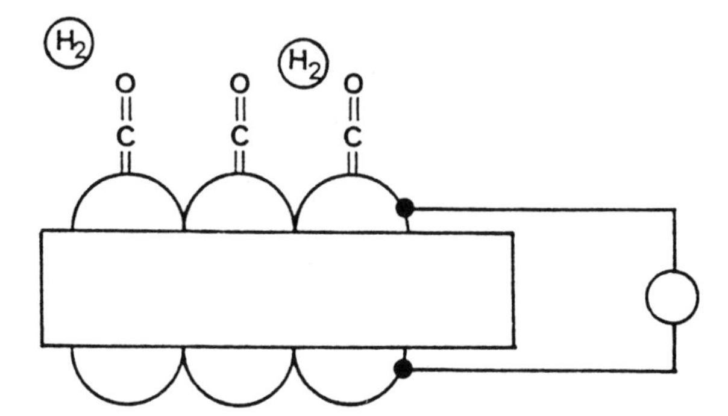

Figure 2.21b Molecular hydrogen cannot reach the active sites on the platinum and dissociate. The electrode is poisoned.

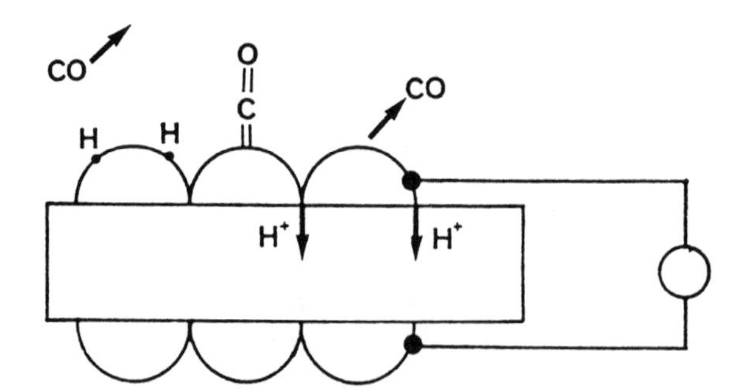

Figure 2.21c When the CO is removed from the gas stream, the CO on the platinum surface desorbs, exposing the active sites for hydrogen adsorption and dissociation.

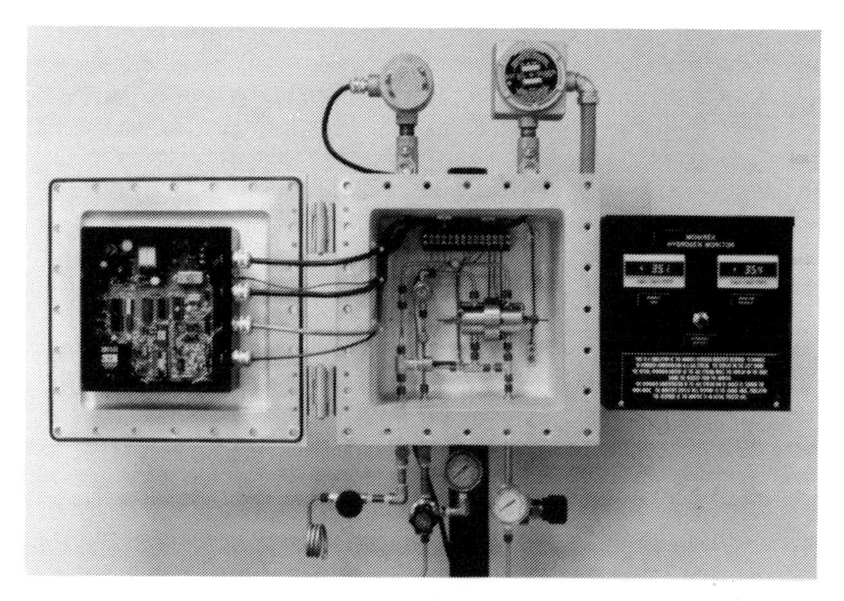

Figure 2.22 UOP Monirex Systems Hydrogen Analyzer.

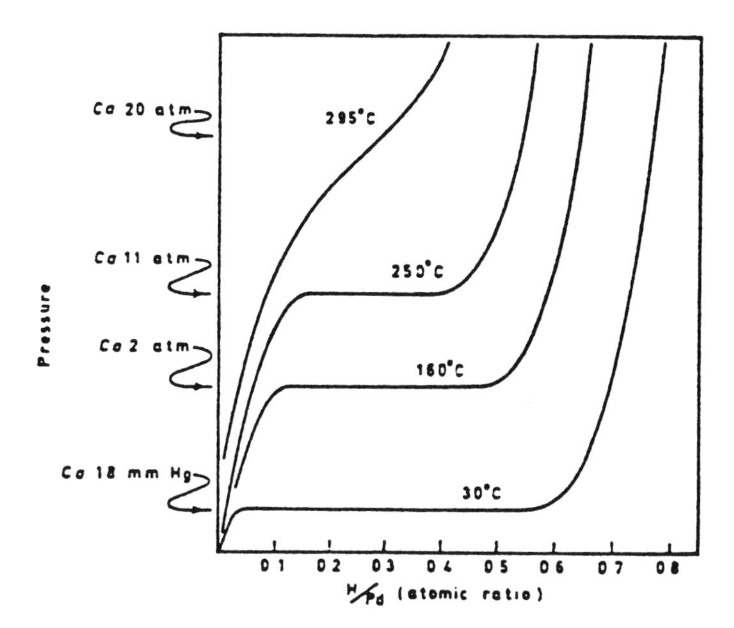

Figure 2.23 Hydrogen partial pressure of palladium hydride complexes as a function of the H-to-Pd ratio.

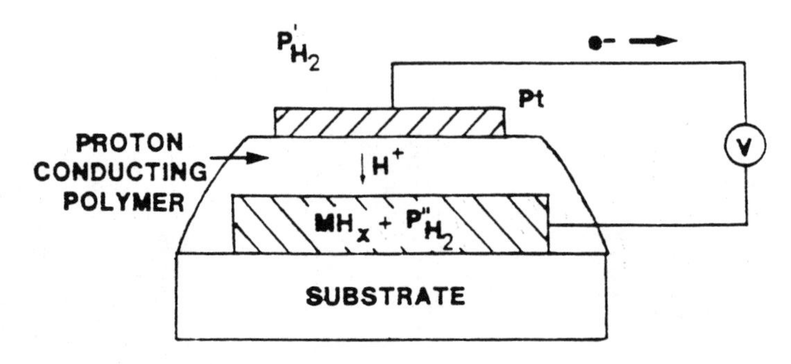

Figure 2.24 Schematic drawing of a solid state hydrogen sensor.

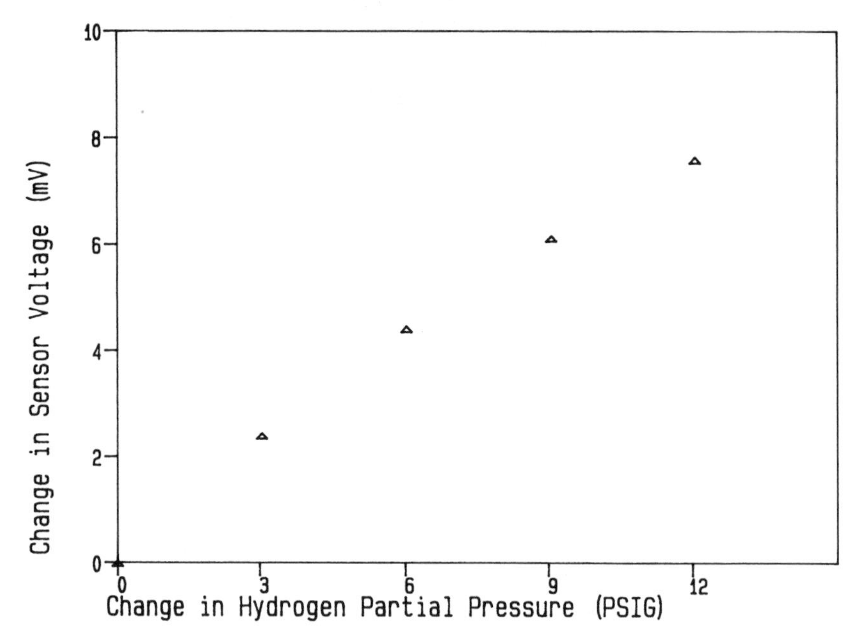

Figure 2.25. Response of a solid state hydrogen sensor to changes in hydrogen pressure (concentration).

material exhibits significant, reversible optical absorption at visible wavelengths.[80] Using this definition, electrochromic materials can be grouped into two categories: inorganic materials that undergo color changes by a double-charge injection process and organic materials that change color by a electroredox reaction. Included in the first group are iridium oxide and tungsten trioxide. The second group includes members of the viologen family of compounds,[75,76] prussian blue and the diphthalocyanines (and rare earth diphthalocyanines[77]). The most thoroughly studied electrochromic materials are the transition metal oxides,[78,79] in particular WO_3. The actual mechanism by which the color change occurs is not well understood, but is generally believed to involve a double-charge injection mechanism (the injection of an electron and a counterion into the electrochromic material). To induce a color change in an electrochromic material (such as WO_3, which is initially transparent), an electron is injected into the electrochromic film, forming the blue tungsten bronze, H_xWO_3. To ensure charge neutrality, a corresponding counterion, such as H^+, must also be injected into the film. It has been found that the resulting change in color, transparent to blue, does not depend on the nature or charge of the counterion. This suggests that the color change is due to the interaction of the injected electron and the WO_3 lattice. To take advantage of the electrochromic effect to form a display, a method of reversibly injecting electrons and ions into an electrochromic film must be devised. One method is to use an ionic conducting electrolyte to separate the electrochromic film from an electrode, which is a source of ions. Such an arrangement is shown in Figure 2.26, in which a multilayer structure is formed. The structure is composed of a sheet of glass supporting a thin, transparent electronic conductor such as indium–tin–oxide (ITO), on top of which is deposited a layer of tungsten trioxide, or other electrochromic material. A tungsten trioxide layer, approximately 500–1000 Å thick, is easily deposited by vacuum evaporation or sputter deposition. Electrochromic materials, such as the diphthalocyanines, can be deposited using vacuum sublimation techniques. On top of the electrochromic material, a thin ionic conducting polymer film is cast. Giglia and Haacke[81] and Calvert et al.[82] utilized a family of sulfonic acid polymers for this layer. The highest conductivity that they observed was with polymerized 2-acrylamido-2-methylpropansulfonic acid (poly-AMPS), which has a room-temperature proton conductivity of 5×10^{-6} (ohm-cm)$^{-1}$ (at a relative humidity of 65%).[81] Another polymer system that has been successfully used as the solid electrolyte in an electrochromic display is the proton-conducting polymer blend composed of PVA/H_3PO_4, which has a room-temperature ionic conductivity of 10^{-5} (ohm-cm)$^{-1}$ in a dry gas environment. To obtain a high contrast between the bleached and colored state of the electrochromic material, a white pigment (such as TiO_2 or Al_2O_3 powder) is normally mixed in with the ionic conducting polymer. In an electrochromic display that uses WO_3 as the electrochromic material, the polymer must be a cation conductor, such as PVA/H_3PO_4. Following the application of an ionic conducting polymer film, an electrode composed of a material that acts as a proton source, such as PdH_x or

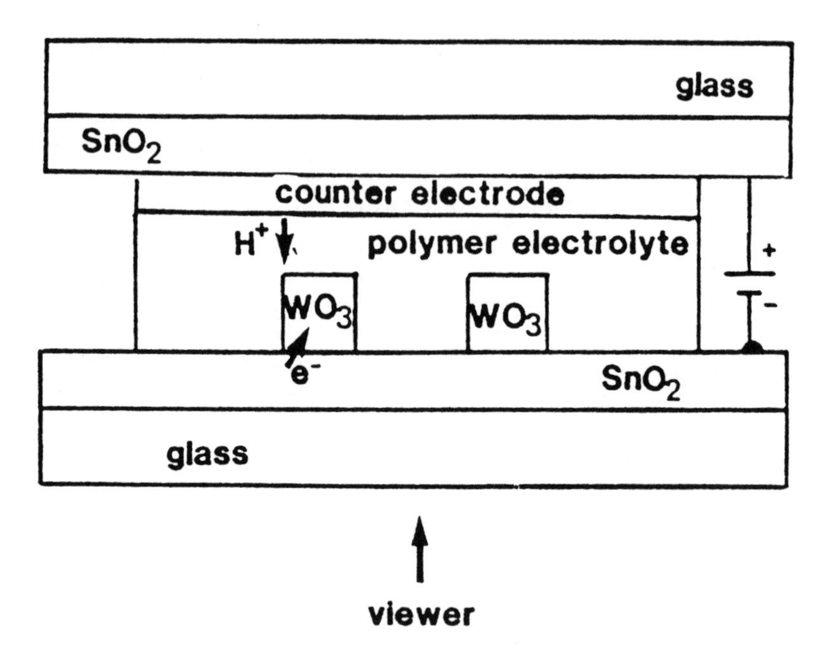

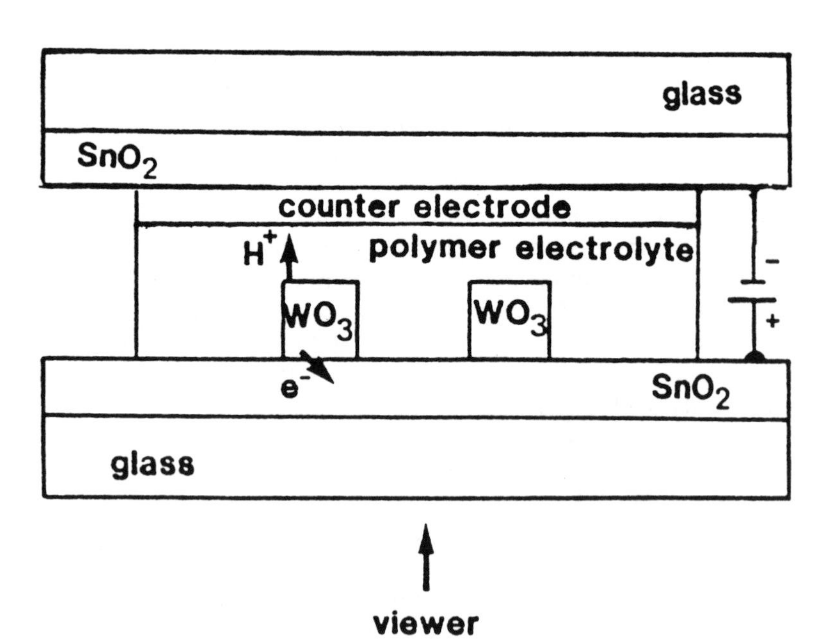

Figure 2.26 Schematic drawing of an electrochromic display, in which the electrochromic layer (WO_3) is in the bleached state (upper), and in the colored state (lower).

H_xWO_3, is deposited on top of the polymer. Electrical connections are now made to both electrodes. When the electrochromic layer is biased positive with respect to the hydrogen electrode, the WO_3 is transparent and all that is seen is the white background of the polymer–pigment mixture. If the polarity is reversed, electrons are injected into the electrochromic layer through the ITO while protons are injected into the electrochromic layer from the hydrogen electrode through the proton conducting polymer film. This results in the formation of the blue hydrogen bronze. This process is easily reversible, and Figure 2.27 shows the response of the electrochromic display described above to a step change in voltage across the electrodes. The response can be seen to be faster in going from the bleached to colored state (about 50 ms) than from the colored to bleached state (about 100 ms). This difference may be due to the differences between the proton diffusion coefficient in the polymer and that within the WO_3 layer. Electrochromic displays offer many potential advantages over LED and LCD displays, such as low power consump-

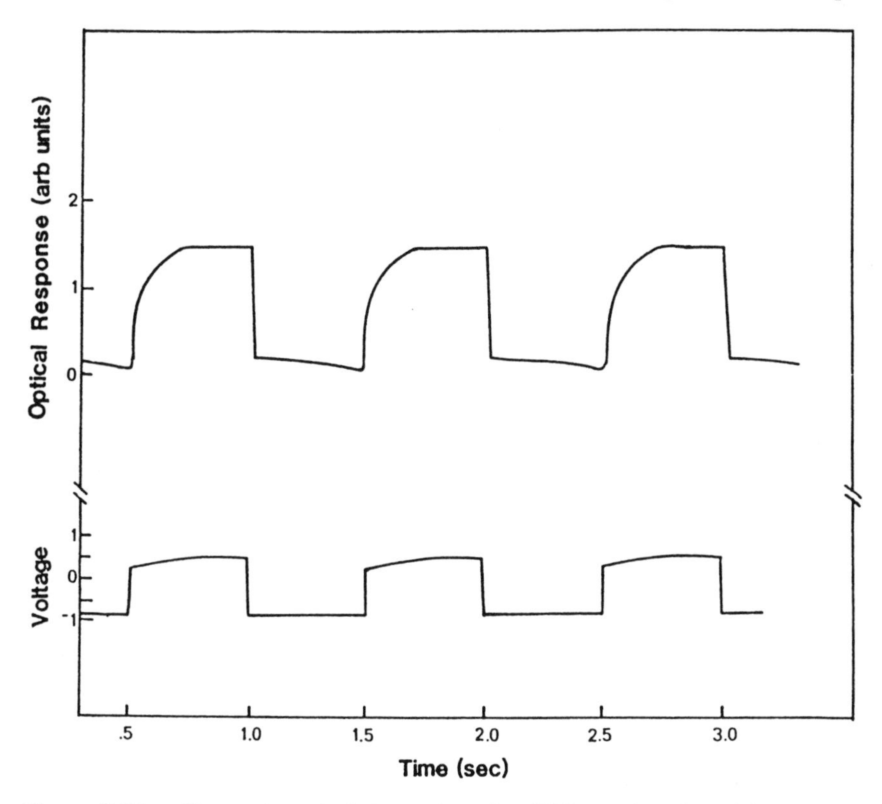

Figure 2.27 Change in optical absorption of an ECD as a function of the applied voltage.

tion (compared to LED displays, but requiring more power than LCDs), multiple colors, and a wider viewing angle than LCDs. The major disadvantage of ECDs has centered on their lifetime, particularly the cycle lifetime of the electrochromic material. Most electrochromic displays are fabricated with electrolytes that either react electrochemically with the electrochromic layer (after several millions of cycles, dissolution of the electrochromic material is observed) or are electrochemically more stable, but result in unacceptably long response times. Whether or not this would be a problem with the ionic conducting polymer electrolytes remains to be determined.

Polymer Electrolyte Photoelectrochemical Cells

Skotheim and Inganas[82] used the ionic conducting polymer PEO complexed with KI and I_2 to generate the iodide–triiodide redox couple, which they then incorporated into a photoelectrochemical cell (PEC), shown in Figure 2.28. Extremely thin films were used (<1000 Å) to compensate for the low conductivity of the PEO complex. At this thickness the polymer films are essentially transparent and allow current densities on the order of milliamperes to be achieved, as shown in Figure 2.29. Although the efficiency of this system was quite low, on the order of a few percent, no corrosion of the silicon was observed. Systems such as this offer the advantage of ease of fabrication and the ruggedness and stability of an all solid-state photovoltaic device.

Polymer Electrolyte-Based Batteries

The fact that liquids as well as solids can support high ionic conductivity has been of critical importance in the development of the first liquid electrolyte battery and the newest generation of solid-state batteries, such as the sodium–sulfur battery, which uses a sodium β-alumina solid electrolyte and must be operated at temperatures greater than 300°C.

With the exception of the sodium–sulfur battery, almost all present-day alkali–metal batteries are lithium-based systems, and the majority of these incorporate either a liquid electrolyte or a liquid cathode. The application of all solid-state cells has been restricted to very limited markets, such as power sources for cardiac pacemakers. Interestingly, the solid electrolyte in the cardiac pacemakers is lithium iodine, a poor lithium ion conductor.

The idea of an all solid state battery is not new.[83,84] Shortly after the discovery of the fast ion conducting properties of sodium β-alumina, Hever,[84] constructed a ceramic rechargeable battery based on a β-alumina solid electrolyte and β-ferrite mixed conducting electrodes. Because of

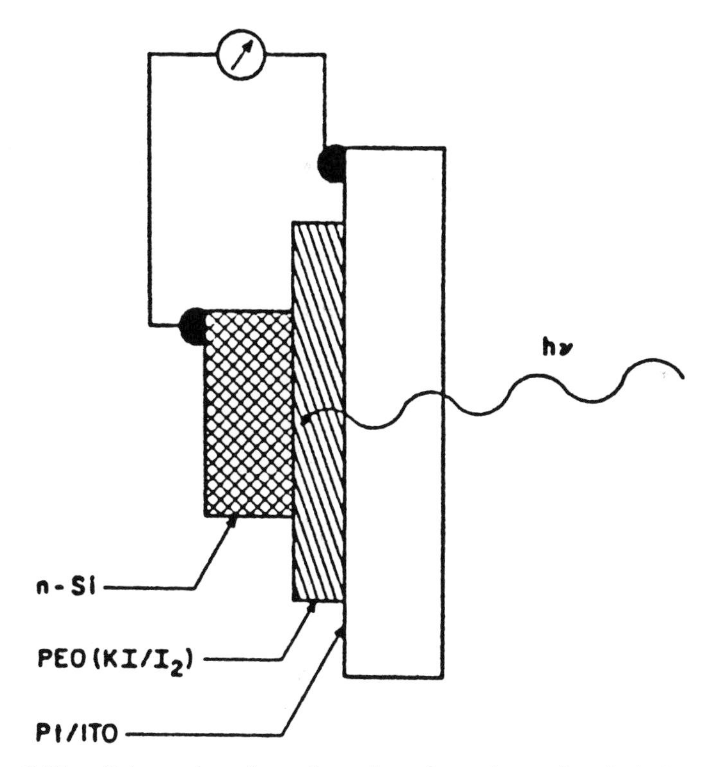

Figure 2.28 Schematic cell configuration of an electrochemical photovoltaic cell.

the device's high internal resistance, it had to be operated at temperatures greater than 300°C and suffered from internal shorting at these temperatures. Mismatches in the coefficient of thermal expansion between the electrolyte and electrodes would probably have resulted in long-term stability problems with such a device. In 1972 the all solid state cardiac pacemaker battery was introduced. The Li/I_2 [poly(2-vinyl pyridine)] pacemaker battery is the only successful commercialization of an all solid-state battery to date. This battery was able to provide a much higher energy density and therefore longer lifetime in use than competing systems, which is the reason for its success. The battery is reliable in delivering years of service at the low current levels required.

Although numerous crystalline inorganic electrolytes are known to have ionic conductivities approaching those of the liquid electrolytes and are used as the solid electrolyte in battery applications, research in the field of solid electrolytes is moving in the direction of developing solid polymer electrolytes.

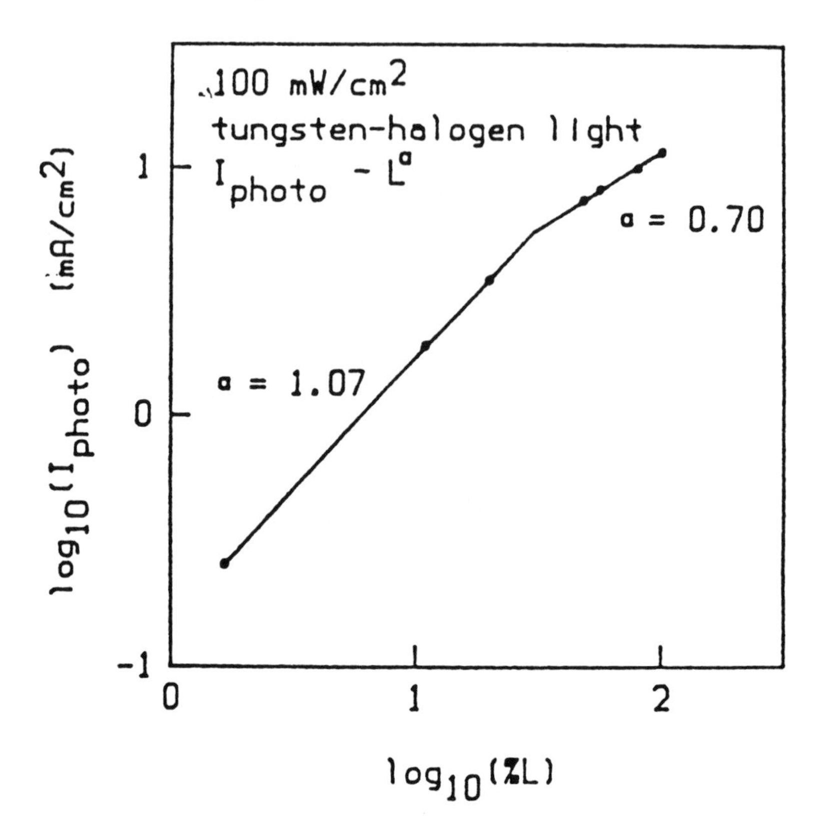

Figure 2.29 Dependence on incident light intensity of the short circuit photocurrent for a cell with n-Si/Pt/polypyrrole iodide photoanode.

The use of ionic conducting polymer electrolytes for battery applications was pioneered by Michel Armand's research group at the University of Grenoble in France. Polymers are attractive as the solid electrolyte in a battery because they

1. Offer the potential for high energy density—up to 200 Wh/kg delivered energy density in high-rate secondary systems.
2. Are easily fabricated.
3. Are flexible and possess good interfacial properties, maintaining contact with the electrodes as the battery is charged and discharged.
4. Have a low self-discharge rate.
5. Can operate over a wide temperature range.
6. Are rugged, since they contain no liquid components.

The biggest drawback in using a solid-state polymer electrolyte is the reduction in conductivity that occurs at temperatures below 80°C, permitting only a modest current to be drawn at room temperature. Progress in developing polymer electrolytes that operate closer to ambient temperature, such as the polyphosphazenes developed by Shriver et al. and the PVA/H_3PO_4 blend developed by Polak et al.,[9] is promising.

Perhaps the biggest advantage that polymer electrolytes offer over the conventional inorganic electrolytes is their ability to deform under stress. This makes them an extremely promising material for use as the solid electrolyte in an all-solid-state battery. The greatest difficulty in developing a solid-state battery arises from dimensional changes that occur in the electrodes during charging and discharging. In a typical high-energy-density battery, lithium might be used as the negative electrode. As the battery discharges, oxidation of the electrode gradually removes lithium from the metal–electrolyte interface. This results in a dimensional change in the lithium electrode and a stress at the electrode–electrolyte interface. If the interface cannot deform to accommodate this stress, contact between the electrode and electrolyte will be lost and the device will cease to function. Similarly, the insertion of Li^+ into the cathode material [which may be V_6O_{13} or titanium disulfide (TiS_2)] causes the cathode to swell, which may result in interfacial delamination at the cathode. Since most ionic conducting polymers have a low modulus (compared with that of the hard crystalline solid electrolytes), they have the potential to maintain contact with the electrodes even after repeated charging cycles.

Another advantage of polymer electrolytes is the ease with which they can be cast into thin films. Thin films minimize the resistance of the electrolyte and reduce the volume and weight of the battery. When thin-film electrolytes are used, a greater fraction of the battery weight and volume can be devoted to the active electrodes, which results in an increase in the amount of energy stored per unit volume or weight.

Examples of the components of a polymer electrolyte battery are shown in Figures 2.30a and 2.30b. In the cell configuration shown in Figure 2.31 the electrolyte is a complex of poly(ethylene oxide) and LiF_3CS_3, the cathode is a composite phase composed of the electrolyte together with V_6O_{13} and acetylene black, and the anode is made of metallic lithium foils. To achieve the high surface area required for high current operations, batteries using both spiral-wound and folded configurations have been developed. Typical cathode specific capacities are in the 2–2.5 mAh/cm^2 range.[85–87]

In addition to having the capability of undergoing numerous charging and discharging cycles, a successful, large, multicell battery must also be able to cycle in both series and parallel connected cell arrays. Hooper et al.[86] have fabricated and tested a three-cell bipolar battery with an effective area of 15 cm^2 at 130°C. The first-cycle behavior is shown in Figure

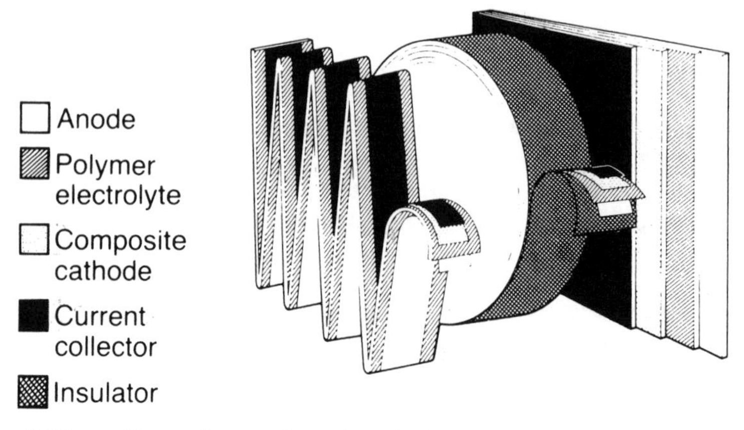

Figure 2.30a Alternative configurations for large area, all solid state cells.

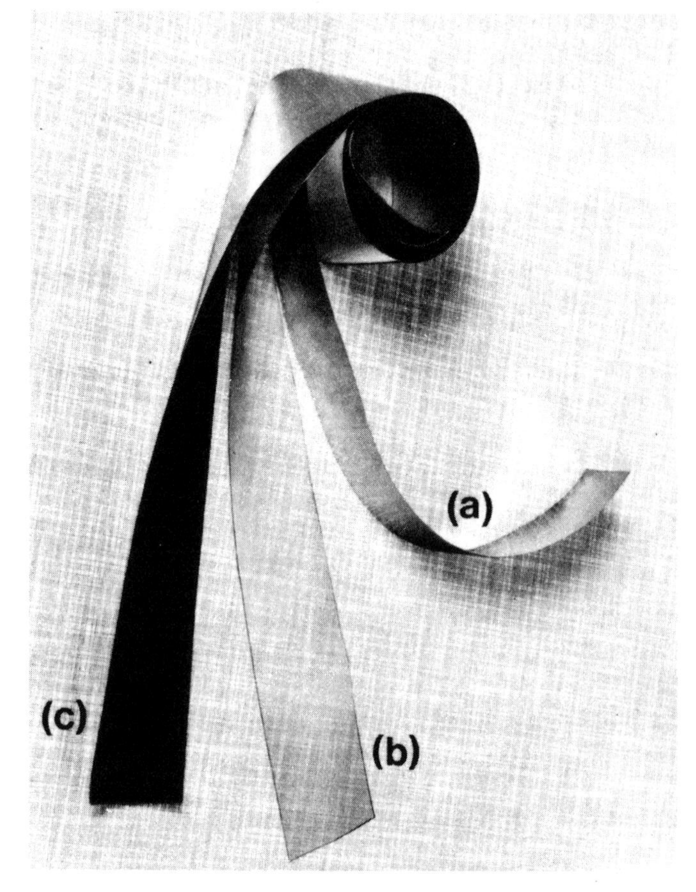

Figure 2.30b Solid state cell components: (a) lithium metal anode; (b) polymer electrolyte membrane; (c) composite cathode on nickel foil backing.

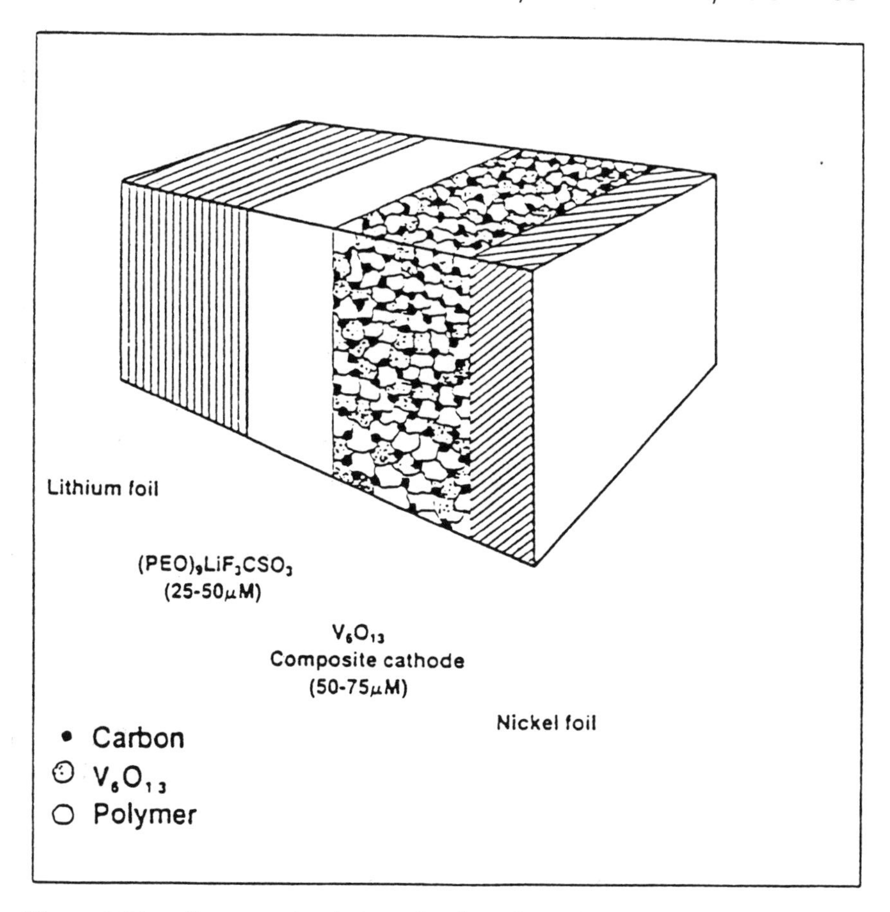

Figure 2.31 Cross-sectional view of cell configuration.

2.32. The discharge and charge currents are 0.2 and 0.1 mA/cm^2, respectively, with voltage limits of 5.1 and 9.75 V. Following 25 cycles, the battery capacity falls from 100% to 55% of the theoretical value (based on $Li_8V_6O_{13}$). By improving the physical structure of the cathode, and running the batteries at lower rates, cell capacities on the order of 50% to 80% of theoretical have been maintained after 35 and 100 cycles, as shown in Figure 2.33.

Although at the present time no batteries based on an ionic conducting polymer electrolyte are commercially available, their potential advantages continue to stimulate a great deal of interest. The unique advantages of polymer-based batteries suggest a wide range of applications. Figure 2.34 shows a qualitative matrix of some of the important advantages of polymer-based batteries for different applications. The major

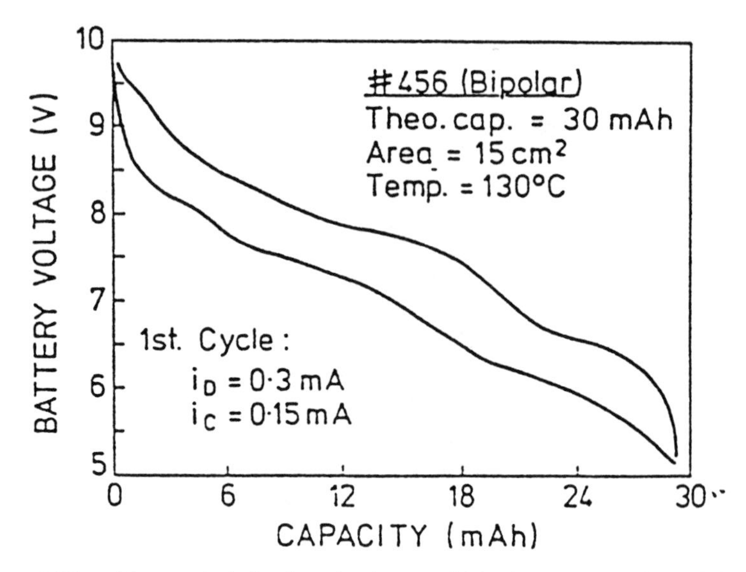

Figure 2.32 First cycle behavior of a three cell bipolar connected, $Li/PEO/V_6O_{13}$ battery.

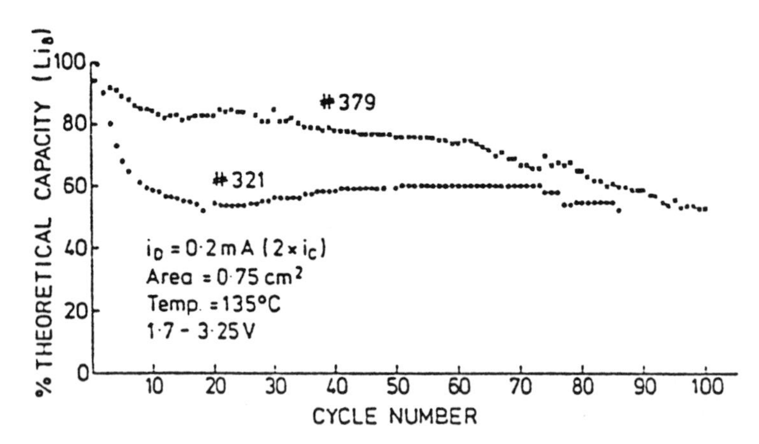

Figure 2.33 Discharge capacity vs. cycle number for two $Li/PEO/V_6O_{13}$ cells, showing improved early life performance.

technical problem that needs to be addressed before polymer-based batteries can become commercially viable is their high operating temperature. However, major applications already exist where above-ambient operating temperatures are acceptable. The high energy densities that can be generated from these systems makes them particularly attractive. The viability of rechargeable solid-state batteries for a range of applica-

	traction	elec-tronics	communica-tions	solar	consumer	military	space
High energy denisity	#	*	*	*	*	#	#
saftey	#	*	*	*	#		
ruggedness	*			*	*	*	#
ease of fabrication	#			*	#		
variable geometry		#	#	#	#	*	
low self-discharge		*	*	#		#	
lack of passivation		*	*	*		*	
requirement for ambient or sub-ambient temp.	not if thermally insulated	yes	yes	yes	yes	yes	no

#-Major Advantage *-Significant Advantage

Figure 2.34 Advantages of all solid state batteries and their application areas.

tions has already been demonstrated on a laboratory scale, and demonstration scale-up programs are now required.

References

[1] G. A. Brown, in *Polymer Materials for Electronic Applications,* American Chemical Society (1982), p. 151.

[2] M. Armand, J. M. Chabagno, and M. J. Duclot, in *Fast Ion Transport in Solids,* P. Vashista, J. N. Mundy, and G. K. Shenoy, eds., North-Holland, New York, 1979, p. 131.

[3] M. B. Armand, J. M. Chabagno, and M. J. Duclot, in *Second International Meeting on Solid Electrolytes,* St. Andrews, Scotland, 1978, Abstract 6.5.

[4] R. G. Linford and S. Hackwood, *Chem. Rev.* 81, 327 (1981).

[5] C. A. Angell, *Solid State Ionics* 18–19, 72 (1986).

[6] J. O'M. Bockris and A. K. N. Reddy, in *Modern Electrochemistry*, Plenum, New York, 1970.

[7] B. E. Fenton, J. M. Parker, P. V. Wright, *Polymer* 14, 589 (1973).

[8] P. M. Blonsky, D. F. Shriver, P. Austin, and H. R. Allcock, *J. Am. Chem. Soc.* 106, 6854 (1984).

[9] A. J. Polak, S. Petty-Weeks, A. J. Beuhler, *Sensors and Actuators*, 9, 1 (1986).

[10] Wetton, *Ions in Polymers*, Advances in Chemistry Series No. 187, A. Eisenberg, ed., Washington, D.C. 1980.

[11] M. Armand, *Solid State Ionics* 9–10, 745 (1983).

[12] P. M. Blonsky, D. F. Shriver, P. Austin, and H. R. Allcock, *J. Am. Chem. Soc.* 106, 6854 (1984).

[13] P. M. Blonsky, S. Clancy, L. C. Hardy, C. S. Harris, R. Spindler, and D. F. Shriver, *Poly. Mater. Sci. Eng.* 53, 736 (1985).

[14] R. Dupon, B. L. Papke, M. A. Ratner, and D. F. Shriver, *J. Electrochem. Soc.* 131, 586 (1984).

[15] R. M. Blonsky, D. F. Shriver, P. Austin, and H. R. Allcock, *Solid State Ionics* 18–19, 258 (1986).

[16] H. Cheradame, J. L. Souquet, J. M. Latour, *Mater. Res. Bull.* 15, 1173 (1980).

[17] J. E. Bauerle, *J. Phys. Chem. Solids* 30, 2657 (1969).

[18] C. K. Chaing, G. T. Davis, and C. A. Harding, *Solid State Ionics* 18–19, 300 (1986).

[19] R. C. T. Slade, A. Hardwick, and P. G. Dickens, *Solid State Ionics* 8–9, 1093 (1983).

[20] L. C. Hardy and D. F. Shriver, *J. Am. Chem. Soc.* 107, 3823 (1985).

[21] P. V. Wright, *Br. Poly. J.* 7, 319 (1975).

[22] P. V. Wright, *J. Poly. Sci. Poly. Phys. Ed.* 14, 955 (1976).

[23] T. M. A. Abrantes, L. J. Alcacer, and C. A. C. Sequeira, *Solid State Ionics* 18–19, 315 (1986).

[24] E. A. Rietman, M. L. Kaplan, and R. J. Cava, *Solid State Ionics* 17, 67 (1985).

[25] J. R. MacCallum, M. J. Smith, and C. A. Vincent, *Solid State Ionics* 11, 307 (1984).

[26] Y. Ito, K. Syakushiro, M. Hiratani, K. Miyauchi, and T. Kudo, *Solid State Ionics* 18–19, 277 (1986).

[27] W. Gorecki, R. Andreani, C. Berthier, M. Armand, M. Mali, J. Roos, D. Brinkmann, *Solid State Ionics* 18–19, 295 (1986).

[28] P. Ferloni, G. Chiodelli, A. Magistris, and M. Sanesi, *Solid State Ionics* 18–19, 265 (1986).

[29] T. Takahashi, G. T. Davis, C. K. Chiang, and C. A. Harding, *Solid State Ionics* 18–19 (1986).

[30] A. Killis, J. F. Le Nest, A. Gandini, H. Cheradame, and J. P. Cohen-Addad, *Solid State Ionics* 14, 231 (1984).

[31] M. L. Williams, R. F. Landel, and J. D. Ferry, *J. Am. Chem. Soc.* 77, 3701 (1955).

[32] C. A. Angell and C. T. Moynihan, in *Molten Salts,* G. Mamantov, ed., Marcel Dekker, New York, 1969, pp. 315–375.

[33] G. S. Fulcher, *J. Am. Ceram. Soc.* 8, 339 (1925).

[34] A. K. Doolittle, *J. Appl. Phys.* 22, 1471 (1951).

[35] T. G. Fox and P. J. Flory, *J. Appl. Phys.* 21, 581 (1950).

[36] N. F. Sheppard and S. D. Senturia, *J. Poly. Sci. Poly. Phys. Ed.,* submitted for publication, 1985.

[37] M. H. Cohen and D. Turnbull, *J. Chem. Phys.* 31, 1164 (1959).

[38] A. Killis, J. F. Le Nest, H. Cheradame, and A. Gandini, *Makromol. Chem.* 183, 2835 (1982).

[39] S. I. Smedley, *The Interpretation of Ionic Conductivity in Liquids,* Plenum, New York, 1980.

[40] G. Adam and J. H. Gibbs, *J. Chem. Phys.* 43, 139 (1965).

[41] J. H. Gibbs and E. A. DiMarzio, *J. Chem. Phys.* 28, 373 (1958).

[42] P. R. Sorensen and T. Jacobsen, *Electrochim.* Acta. 27, 1671 (1982).

[43] J. E. Weston and B. C. H. Steele, *Solid State Ionics,* 2, 347 (1981).

[44] E. Tubandt, *Handb. Experimentalphys.* 12, 383 (1932).

[45] Monirex Systems Hydrogen Analyzer, UOP, Inc., Des Plaines, Ill.

[46] A. Bouridah, F. Dalard, D. Deroo, and M. B. Armand, *Solid State Ionics* 18–19, 287 (1986).

[47] M. Watanabe, M. Rikukawa, K. Sanui, and N. Ogata, *J. Appl. Phys.* 58, 736 (1985).

[48] M. Watanabe, K. Sanui, N. Ogata, T. Kobayashi, and Z. Ohtaki, *J. Appl. Phys.* 57, 123 (1985).

[49] M. Watanabe, S. Nagano, K. Sanui, and N. Ogata, *Solid State Ionics* 18–19, 338 (1986).

[50] J. P. Stagg, *Appl. Phys. Lett.* 31, 532 (1977).

[51] G. Greeuw and J. F. Verwey, *J. Appl. Phys.* 56, 2218 (1984).

[52] G. Greeuw and B. J. Hoenders, *J. Appl. Phys.* 56, 3371 (1984).

[53] M. Kosaki, H. Ohshima, and M. Ieda, *J. Phys. Soc. Jpn.* 29, 1012 (1970).

[54] C. Berthier, W. Gorecki, M. Minier, M. B. Armand, J. M. Chabagno, and P. Rigaud, *Solid State Ionics* 11, 91 (1983).

[55] S. Bhattacharja, S. W. Smoot, and D. H. Whitmore, *Solid State Ionics* 18–19, 306 (1986).

[56] C. Berthier, Y. Chabre, W. Gorecki, P. Segransan, and M. B. Armand, in *Proceedings of the 160th Meeting of the AECS,* Denver, Vol. 81–82, 1981, p. 1495.

[57] S. Petty-Weeks and A. J. Polak, *Sensors and Actuators* 11, 377 (1987).

[58] S. G. Cutler, in *Ions in Polymers,* Advances in Chemistry Series, A. Eisenberg, ed., Vol. 187, Washington, D.C., 1980.

[59] T. D. Gierke, 152nd National Meeting Electrochemical Society, Atlanta, Ga., 1977.

[60] A. J. Hopfinger and K. Mauritz, in *Comprehensive Treatise of Electrochemistry,*

Vol. 2, J. M. Brockris, B. E. Conway, E. Yeager, and R. E. White, eds., Plenum, New York, 1981.

[61] R. Wodzki, A. Narebska, and W. K. Nioch, *J. Appl. Poly. Sci.* 30, 769 (1985).

[62] W. Y. Hsu, J. R. Barkley, and P. Meakin, *Macromolecules,* 13, 198 (1980).

[63] W. Y. Hsu and T. Berzins, *J. Poly. Sci. Poly. Phys. Ed.* 23, 933 (1985).

[64] S. D. Druger, M. A. Ratner, and A. Nitzan, *Phys. Rev B* 31, 3939 (1984).

[65] S. D. Druger, A. Nitzan, and M. A. Ratner, *J. Chem Phys.* 79, 3133 (1983).

[66] S. D. Druger, M. A. Ratner, and A. Nitzan, *Solid State Ionics* 9–10, 1115 (1983).

[67] O. Worz and H. R. Cole, *J. Chem Phys.* 51, 1546 (1969).

[68] M. G. Shilton and A. T. Howe, *Mater. Res. Bull.* 12, 701 (1977).

[69] L. Bernard, A. N. Fitch, A. F. Wright, B. E. F. Fender, and A. T. Howe, *Solid State Ionics* 5, 459 (1981).

[70] K. Kiukkola, C. Wagner, *J. Electrochem. Soc.* 104, 379 (1957).

[71] E. Baun and H. Pieis, *Z. Electrochem.* 43, 727 (1937).

[72] F. Haber, *Z. Anorg. Chem.* 57, 154 (1908).

[73] W. P. Treadwell, *Z. Electrochem.* 22, 411 (1916).

[74] F. A. Lewis, *The Palladium Hydrogen System,* Academic Press, New York, 1967.

[75] P. N. Moskalev and I. S. Kirin, *Opt Spectroscosc.* 29, 220 (1970).

[76] C. J. Scoot, J. J. Ponjee, H. T. van Dam, R. A. van Doorn, and P. T. Bolwijn, *Appl. Phys. Lett.* 23, 64 (1973).

[77] M. M. Nicholson, *Society for Information Display Digest,* February 1984, p. 1.

[78] W. C. Dautremont-Smith, *Displays,* January 1982, p. 3.

[79] I. F. Chang, Proc. Brown Boveri Symposia, *Nonemissive Electrooptic Displays,* A. R. Kmetz and F. K. von Willisen, eds., 1975, pp. 155–196.

[80] T. Oi, *Annual Review of Materials Science,* Vol. 16, R. A. Huggins, J. A. Giordmaine, and J. B. Wachtman, eds., 1986.

[81] R. D. Giglia and G. Haacke, *SID Dig.* 12, 76 (1981).

[82] J. M. Calvert, T. J. Manuccia, and R. J. Nowak, *J. Electrochem. Soc.* 133, 591 (1986).

[82] T. A. Skotheim and O. Inganas, *J. Electrochem. Soc.* 132, 2116 (1985).

[83] B. C. Tofield, NATO Advanced Study Institute on Solid State Batteries, Portugal (1984), "Future Prospects for All-Solid-State Batteries" AERE-R 11576, AERE Harwell.

[84] K. O. Hever, *J. Electrochem. Soc.* 115, 830 (1968).

[85] A. Hooper and J. M. North, *Solid State Ionics* 9–10, 1161 (1983).

[86] A. Hooper, J. M. North, M. Oliver, and B. C. Tofield, 31st Power Sources Symposium Proceedings, Cherry Hill, N.J., 1984.

[87] A. Hooper, private communication.

[88] W. Y. Hsu, J. R. Barkley, and P. Meakin, *Macromolecules* 13, 198 (1980).

[89] W. Y. Hsu and T. Berzins, *J. Poly. Sci. Poly. Phys. Ed.* 23, 933 (1985).

Part II

Plastics

Chapter 3

Metallic Plating and Coating on Plastics

James M. Margolis

Introduction

Metallic plated and coated plastic products are used primarily for decorative applications, automotive and appliance products, plumbing and marine applications, and other consumer and industrial products. On a tonnage basis 20% to 25% of metallized plastics are used for conductive applications. Conductive plastic products are used to provide electrostatic discharge (ESD) and electromagnetic interference–radiofrequency interference (EMI/RFI) shielding.

All EMI/RFI applications are related to electrical devices, and most ESD applications are related to electronic applications. Plastic fluid-handling equipment, textile bobbins, and any application where friction occurs on the surface of the plastic will generate electrostatic charge.

Table 3.1 shows consumption of metallic plated and coated plastics for electrical and electronic applications as compared with other applications in the United States for 1985 and 1990.

Any plastic can be plated, but specific resin grades must be used for plating. Two principal characteristics of platable and coatable grades are (1) the surface of the finished part must form a strong bond with the metallic layer (i.e., there must be good adhesion between the plastic substrate and metallic layer), and (2) resin specifications must be highly uniform from shipment to shipment.

Two other criteria influence the use of different plastics: (1) certain plastics are inherently more easily platable than others and (2) certain plastics have other features, such as lower cost and mechanical proper-

Table 3.1 Consumption of Metallic Plated and Coated Plastics (in millions of pounds)

	1985	1990
Electrical–electronic	25	45
Other industries	117	145
Total	142	190

ties, that are suitable for specific applications. (Remember that most applications are for decorative purposes). ABS resins meet metallizing characteristics and criteria better than other plastics, and about 50% of all metallized plastics are ABS. However, the percent of total ABS is decreasing as engineering thermoplastics are increasingly being used for metallized applications.

The principal plastics used for electrical and electronic applications and other applications that require ESD are ABS, nylon (solid and foam), polycarbonate, ABS–polycarbonate alloys, polystyrene (foam), acrylics, acetal, vinyls, polypropylene, urethane (plastic and elastomer) polyphenylene ether–polyphenylene oxide (solid and foam), polysulfone, polyethersulfone, polyphenylene sulfide, polyesters (thermoplastic and thermosetting), and epoxy (thermosetting). Epoxy resins are the most-used resins for printed circuit boards at present.

Most metallized plastics products are made by electroplating, with the use of electroless plating increasing for EMI/RFI applications. The other principal processes are vacuum metallizing, sputtercoating (a vacuum process), arc spray and flame spray, and hot foil stamping. Electroplating, including electroless plating, accounts for 80% of all metallized coated plastics.

Plastics

New platable plastic grades are being introduced, especially engineering thermoplastics with flame retardant values of UL94 V/O-5V, high-temperature plastics that retain a significant level of service properties at 200°C, and plastics with improved mechanical properties. The trend to higher-performance metallized plastics is due to high-tech applications, especially electrical and electronic applications.

Three fundamental properties important for metallizable plastics are (1) coefficient of linear thermal expansion, C_e, (2) deflection temperature under load (DTUL), and (3) strength.

The principal cause of blistering and cracking is different coefficients of linear thermal expansion between the metallic surface and the plastic substrate. The different C_e between the metal film and the plastic substrate creates a stress at the interface between the film and substrate. The typical C_e for nickel is 0.7×10^{-5}, and for copper it is 0.9×10^{-5} in.in./°F. DTUL provides dimensional stability to the molded part in hot electroplating baths. A representative value is 235°F/264 psi for platable-grade polyphenylene oxide (PPO).

DTUL and strength and moduli mechanical properties value requirements depend on the end product application. A comparison of the mechanical properties of platable-grade ABS before and after plating is shown in Table 3.2. Total plate thickness is 0.0015 in. (0.00006 mm). Recommended electroplate thicknesses for the series of strike (plating film layers) are shown in Table 3.3. Platable-grade PPO was developed to provide increased dimensional stability to plated plastics and to provide better consistency when the materials are exposed to temperature extremes. PPO has low creep, good low-temperature impact resistance, and DTUL 235°F/264 psi, as noted previously. Platable PPO and sputter-coatable PPO can be molded into road vehicle and appliance parts such as passenger car grills, wheel covers, and headlamp bezels and appliance knobs, door bezels, and decorative trim.

Mineral-reinforced nylon can be used for ESD applications such as textile bobbins, but nylon is not used in applications involving continuous contact with water, because of the hydrophilic nature of nylon. Nylon can absorb over 1% (weight) water (24-hr immersion at room temperature) compared with values as low as 0.04% for some resins. Water absorption affects plating-cycle times. Cycles over 30 min can

Table 3.2 Comparison of Cycolac EP-3510 Before and After Electroplating*

	73°F (23°C)	
Mechanical Property	Before Plating	After Plating
Tensile strength		
psi	5,500	6,600
MPa	38	45
Tensile modulus		
psi	330	630
MPa	2,300	4,340
Flexural strength		
psi	10,000	12,000
MPa	69	83
Flexural modulus		
psi $\times 10^3$	340	1,100
MPa	2,300	7,580

* Test specimens were injection-molded 1/8 in. thick bars, tested according to American Society for Testing and Materials (ASTM) methods.

Table 3.3 Recommended Electroplate Thickness

Service Conditions		Recommended Thickness (in.)
	Nickel strike	Adequate to cover
Mild: Exposure indoors in normally warm dry atmosphere	Bright acid copper	0.0006 –0.0008
	Bright nickel	0.0002 –0.0003
	Conventional chromium	0.000010–0.000015
	Nickel strike	Adequate to cover
Moderate: Exposure to high humidity and mildly corrosive atmosphere	Bright acid copper	0.0006 –0.0008
	Semibright nickel	0.0003 –0.0004
	Bright nickel	0.0002 –0.0003
	Conventional chromium	0.000010–0.000015
Severe: Exposure to high humidity, wide temperature variations, and severe corrosive atmosphere	Nickel strike	Adequate to cover
	Bright acid copper	0.0006 –0.0008
	Semibright nickel	0.0004 –0.0006
	Bright nickel	0.0003 –0.0004
	Special nickel*	0.0001
	Conventional chromium	0.000010–0.000015

* Required for inducing microporosity or microcracking.

cause blistering and poor adhesion between the metal film and the plastic substrate. Total preplate cycle time is 18–22 min for either copper or nickel. Plating and metallic coating processes are described later in this chapter. Certain resin grades are developed for either electroplating or electroless plating, whereas other grades can be used for both of these plating methods. Allied-Signal, Inc., Capron CPN-1030 nylon 6 was developed for both electroplating and electroless plating. Nylon generally does not plate easily, but nylon resin manufacturers have introduced improved plating grades since the mid-1980s.

The plating problem involves the wet side, which requires different preplate systems from other resins. Copper preplating improves adhesion, but moisture on the plastic substrate surface and under the metal film causes adhesion problems.

Polycarbonate is the one of the toughest platable plastics, which translates into high impact strength values up to 20 ft lb/in. notch Izod impact. Polycarbonate resins can be plated by modified ABS plating systems, and the resins can be vacuum-metallized, spray-coated and hot-stamped. However, vacuum-metallizing processes do not result in outstanding adhesion between the metal film and plastic substrate, and certain precautions and limitations are required.

Acetals are used for decorative applications, particularly plumbing and marine products. These resins have been difficult to plate, primarily because of the resins' resistance to conventional etch systems. Etching chemical solutions that are effective with acetals can be too strong and tend to cause the acetal to dissolve. Platable grades have been introduced and "reintroduced" since the 1970s, and in the 1980s commercial platable homo- and copolymer acetals were finally commercialized. All steps in the Dupont Delrin homopolymer acetal plating process except the etch are compatible with standard ABS plating processes.

Delrin acetals can be vacuum-metallized following a satinizing process developed by Dupont. Satinizing is a chemical etching process in which a mildly acidic solution is applied to the plastic substrate to form anchor points on the substrate surface. The uniformly distributed anchor points can form strong adhesive bonds to basecoats used in vacuum metallizing, as well as to paints and cements (glues).

Platable high-temperature resins such as polyphenylene sulfides and polysulfones (intermediate high temperature), polyethersulfones, ketone polymers, polyamide-imides, polyetherimides, and polyimides have been produced for ESD and EMI/RFI shielding electrical and electronic applications. These resins not only retain service-level properties such as strength and modulus at temperatures up to 400°F (and above for polyimides), they are inherently flame retardant. That is, flame-retardant chemicals do not have to be compounded into these high-temperature resins to achieve UL 94 V-O flammability ratings.

Plated polyphenylene sulfide (PPS) is flame-spray metal-coated for EMI shielding and military applications, and the resin is vacuum-metallized for electronic circuitry applications. Platable PPS grades with improved metal film peel strength were introduced in the 1980s. Platable polyethersulfone (PES) is used for printed circuits for large telecommunications and computer equipment. During the 1980s injection-molded three-dimensional (3-D) printed circuits were developed, which use PES and other high-temperature plastics. Injection-molded plated thermoplastic printed circuits are an alternative to conventional epoxy 4R laminated printed circuits.

Vapor-phase welding and wave soldering are high-speed, cost-effective assembly techniques for high-volume production of electronic plastic components such as circuit boards, connectors, and chip holders. These assembly techniques require resins with deflection temperatures under load (DTUL at 264 psi) higher than 400°F. Therefore, only thermoplastics with high DTUL values can be used with vapor-phase welding and wave-soldering assembly.

Topcoats and Basecoats

Vacuum metallizing requires lacquer topcoats and basecoats to be applied to the plastic substrate surface. Vacuum-metallized film thicknesses are usually on the order of one-millionth of an inch. The topcoat provides resistance to abrasion and to chemicals. The basecoat provides a compatible, smooth transphase and good adhesion between the plastic substrate surface and the metallic film. Topcoat and basecoat chemical formulations are a highly active area of research in the 1980s for both evaporative and sputter vacuum metallizing. Most lacquers have been urethanes, but ultraviolet (uv) coatings are being used increasingly. Evaporative vacuum coating uses only pure metal, usually aluminum, whereas sputtercoating can use either pure metal or alloys, especially chrome and chrome alloys.

The metal film composition is a factor in topcoat–basecoat selection. The topcoat, such as urethane lacquer, is applied over the thin aluminum film in evaporative vacuum metallizing to provide resistance to wear and abrasion, chemicals, bulk packing, and atmospheric contaminants. The topcoat is usually clear (transparent), but it can be tinted with pigments or dyes to impart copper, gold, brass, or other metallic tones. The topcoat can be flattened to provide a subdued brushed or satin appearance. Overlays are pigmented coatings used for a decorative appearance. For consumer products, decorative surfaces can be used in conductive applications when the conductive surface is visible.

The basecoat is usually a clear, high-build film, such as a urethane lacquer formulated to bond the substrate surface. The basecoat provides a strong bond with the metallic film, which must form a strong bond with the topcoat.

Basecoats and topcoats are often spray-applied using "ride and rod" machines or conveyor chains to rotate parts in front of the spray guns. When ride and rod vehicles are used, the rods must be transferred to a cart that is wheeled into a force-dry oven. When conveyor lines are used, the parts are cured on the conveyor lines in self-contained force-dry ovens. Typical cure times and temperatures are shown in Table 3.4.

Ultraviolet-curable coatings such as cross-linkable Red Spot UVT topcoats and UVB basecoats produced by Red Spot Paint & Varnish Company require only 15 seconds to cure. UVB-9 is an ultraviolet-curable unsaturated acrylic basecoat; and ET 4R4 is an acrylic topcoat. DeSoto, Inc., produces Desolite 370-30A uv-curable topcoats. Uv-curable topcoats

Table 3.4 Typical Cure Times and Temperatures for Several Plastics

Plastic	Oven Temperature (°F)	Cure Time (min)
Polystyrene	140	20–60
ABS	165–175	30–90
Acrylic	165	60–120
Polycarbonate	45	260

and basecoats offer the following advantages over conventional oven-cured coatings:

Energy conservation
Air pollution reduction
Reduction of distortion of parts (from oven heat)
Higher film buildup due to higher solids
Improved film properties due to cross-linking
Reduced equipment costs: $30,000 versus more than $100,000

Uv basecoats are replacing oven-cured varnish basecoats because of the significantly reduced curing time and improved performance properties of the entire metallized coating system. Uv basecoats are superior to conventional basecoats for specific chemical resistance to ketones, alcohols, and essential oils, because of the cross-linking of the uv coatings. However, uv-curable topcoats have shown poor adhesion to aluminum films when exposed to cosmetic chemicals, so for the decorative container caps for these products, force-dried lacquer topcoats are used.

Typical oven-cured topcoats and basecoats for vacuum-metallized plastics are represented by Bee Chemical Company chemicals. Basecoat M-472C is applied by mixing the company's accelerator C-21933 to a 1.5% (volume) accelerator content. The mixture is thinned for either flow coating or spray coating onto the plastic substrate surface prior to metallizing. A typical thinning ratio is one part basecoat–accelerator to three parts thinner (such as T-22108) for flow coating. The flow-coating viscosity is reduced to 33–38 seconds/#1 Zahn. For spray coating the ratio is one part basecoat–accelerator to one part thinner. Specific formulations and ratios are determined by the individual application such as automotive trim or conductive application.

The spray-applied topcoat is not the final step in the metallizing process. The metallized plastic is cured at 165°F bake oven temperature. For maximum performance, especially for abrasion resistance, additional curing at room temperature is carried out for 7 days. The topcoat can be overlaid with additional coatings, such as Bee Chemical coatings NA50, Q-106, and W7E. These overlays are spray- or dip-coated, after thinning.

Design Considerations

Design considerations for metallized plated or coated plastics involve the following:

1. Wall thickness and flatness
2. Radii for corners, edges, ribs
3. Recesses and sink marks
4. Aspect ratio for holes

The performance of metallized plastics is determined by part design as well as by other factors such as materials and metallizing process selections. Basic design principles for molded products apply to plated and coated plastics, and the specific design requirements for individual resins are available from resin suppliers. However, metallized surfaces accentuate surface defects and affect the efficiency of the metallizing process. Therefore, design is more critical for metallized parts than for nonmetallized parts.

Thicker walls usually produce higher peel strength. Peel strength is also affected by resin melt temperatures and dwell time for injection-molded parts. Radii can be expected to be up to 0.5° greater for metallized plastics, especially plated plastics, than the recommended radii for any given plastic that is not metallized. Recesses and sink marks are related to radii or protrusions on the wall, such as ribs and bosses. Aspect ratios (hole length to diameter ratios) for both "blind" holes and through-holes can be expected to be slightly less for metallized plastics, to allow complete and uniform metal plating or coating coverage through the hole.

Metal Plating and Coating Processes

The principal processes for metallized finished plastics parts are (1) electroplating and electroless plating, (2) evaporative vacuum metallizing, (3) vacuum sputtercoating, and (4) hot stamping.

Electroplating

Approximately half of the plastics electroplated in the United States are used for automotive decorative products, and more than 25% are used for decorative appliance products. Electrical and electronic applications, especially electromagnetic interference shielding and copper-plated circuit boards, are the most active electroless plating areas.

A typical electroplating process begins with (1) etching, (2) neutralization, (3) activation, and (4) acceleration. These are the four initial steps prior to plating.

Etching produces a hydrophilic surface and creates sites for adhesion of the metal film deposit. Most commercial etching solutions consist of chromic acid or chromic acid plus sulfuric acid. Neutralization reduces hexavalent chromium (Cr^{6+}) to trivalent chromium (Cr^{3+}), because hexavalent chromium is detrimental to the preplating solutions. Activation "seeds" an ionic palladium catalyst onto the bonding sites that are created by the etching step. Acceleration removes noncatalytic ionic components and exposes the active catalytic sites for the next steps.

Etching creates both mechanical bonds and chemical bonds on the plastic substrate surface. The mechanical bonding is created by etching material out of the surface, which forms fingerlike attachments between the plastic surface and the electroplated metal, usually copper. According to this concept of mechanical bonding, metal is chemically deposited in the pores produced during etching, to form a "stud" anchored in the plastic. The more effective the formation of the pores during etching, the less tendency there is for blistering of the metal film deposit.

The tendency for blistering and peeling in the hot, wet, corrosive plating bath depends on the porosity of the plastic surface when nickel is the initial deposit. However, peel strength actually increases after several months, because of the chemical bonding during etching. Long-term failure of the plated part after several months is due to failures within the plastic part, and not to the etching process. Most electroplated plastics rely primarily on mechanical bonding, which provides stronger bonding than chemical bonds. Chemical bond failure is due to chemical changes at the plastic–metal interface, caused by moisture or the formation of hydrogen ions.

Chemical bonding occurs at low etch-produced porosity (low-profile pore sites) on the plastic surface. Nickel is subject to bond failure due to moisture, but when electroless copper is used instead of nickel, this type of failure does not occur. When ABS is etched with chromic acid, almost no porosity occurs; however, when the etched ABS is first coated with chemical nickel and then electroplated with copper, adhesion between the plastic and metal film improves with air-aging. Bond strength can improve after 24–48 hours.

Chemical bonding increases adhesion after a chromic acid etch on polypropylene with chemical copper when the polypropylene–copper system is aged in oxygen. Adhesion is also increased with etched polypropylene plus chemical nickel even when no porosity develops. To accomplish strong bonding between polypropylene and a nickel coating, the polypropylene is pretreated with turpentine or other organic chemical emulsions. Untreated polypropylene and polyolefins in general have relatively inert surfaces chemically. The organic emulsions dissolve into the polypropylene plastic, creating a surface that is more chemically reactive with chromic acid.

The depth and density of porosity developed during the etching step depends on (1) etch composition, (2) etching time, (3) etching temperature, and (4) type of plastic and design of the molded finished part.

Aqueous colloidal stannous–palladium (Sn/Pd) activators are the most frequently used catalyst systems during the etching step. Stannous ions (Sn^{2+})–hydrochloric acid is used in special applications, followed by immersion in palladium chloride ($PdCl_2$). In one process the palladium is added directly to the chromic acid bath. Strong acids or bases are used as accelerators to activate the catalyst system prior to the deposition of chemical nickel or copper.

The plastic part is immersed in a chemical metal-plating bath after the catalyst is attached to the plastic surface. Metal deposition begins at discrete catalyst sites, forming metal "islands" on the plastic surface. The metal islands further catalyze the metal deposition process until a continuous, electrically conductive metal coating develops. The metal nickel or copper coating is usually 0.25–0.50 μm thick. The initial deposit is from a Watts nickel or copper pyrophosphate strike solution, to protect the relatively thin electroless metal deposit from chemical attack and from the higher voltage in the subsequent electroplating solutions. Without this protection the electroless deposit can partially dissolve or "burn off" at the contact points between the rack and the plastic part. A Watts nickel bath is shown in Table 3.5.

Nickel sulfate provides an inexpensive source for nickel ions, while nickel chloride provides a chloride ion source for suitable anode corro-

Table 3.5 Watts Nickel Plating Solution

Component	Amount	
	(g/liter)	(oz/gal)
Nickel sulfate, $NiSO_4 \cdot 6H_2O$	150–450	(20–60)
Nickel chloride, $NiC1_2 \cdot 6H_2O$	37.5–150	(5–20)
Nickel equivalent as metal Ni	43.–137	(5.7–18.3)
Boric acid, H_3BO_3	37.5–56.3	(5–7.5)
pH	3.5	(Acid)
Temperature	50–79°C	(120–174°)
Cathode current density	108–1080 A/m^2	10–100 A/ft^2
Anode current density	Less than 216 A/m^2	Less than 20 A/ft^2
Agitation of bath	As required	

sion. Without chloride ion–induced corrosion, many anode forms of nickel will not corrode properly, resulting in polarization of the anode and depletion of nickel content from the electrolytic solution. Chloride ions increase conductivity and limiting-current density of the solution. Chloride also has a hardening effect on the deposit, which can reduce ductility of the plated part with certain additional agents. Nickel metal equivalent content of the electrolytic solution is obtained primarily from nickel sulfate and secondarily from nickel chloride. Nickel metal equivalent content affects the limiting cathode current density, which is the electrical current density (amperes per square area) at which the deposit will begin to "burn." Higher nickel ion concentration in bulk solution allows more nickel ions to enter the cathode film, to replenish nickel ions consumed during electrolytic plating.

However, it is best to use lower nickel content to minimize waste disposal. Nickel ion content can be reduced to a minimum content and maximum efficiency level by considering the part design configuration and balancing plating-bath temperature and agitation in the plating solution bath. Increased temperature (up to a point) and agitation can increase the diffusion rate of nickel ions to the cathode film.

Boric acid is the conventional buffering agent in Watts nickel-plating solution. It shows minimum side effects during electroplating, reduces the tendency for gas pitting, and increases limiting cathode density. Boric acid contributes hydrogen ion (H^+) to the cathode film, maintaining the pH below the point where nickel precipitates. The primary effect of boric acid is to buffer the cathode film, which is a thin film comprised of partially depleted nickel solution. Without the buffering effect of boric acid, by its contribution of additional H^+ ions to the cathode film, pH would continue to increase and nickel hydrate would precipitate at high current-density areas. Nickel hydrate is a green salt formed during burning.

High throw acid copper strikes are an alternative to Watts nickel or copper pyrophosphate strike solutions. According to Udylite/Oxy Metal Industries, Inc., high throw acid copper strikes provide better metal distribution, decrease cost, and reduce waste treatment requirements. High throw copper strikes are composed of 10–12 oz/gal copper sulfate plus 24–30 oz/gal sulfuric acid, and 40–100 parts per million (ppm) chloride ions plus proprietary brighteners for decorative applications. The high concentration of sulfuric acid provides good throwing power. Brighteners increase the high current-density ranges (which prevents "burnoff"), increase throwing power, and prevent grain refinement.

High throw acid copper strike solutions are similar to formulations used in plating electronic components. The copper metal deposited in these solutions can be included in the total copper thickness requirement to meet end product specifications. All rinsing between the copper strike and copper plating is eliminated. Only an acid rinse before the strike is needed to ensure an active copper surface. Waste treatment is reduced, and all salts dragged out from the strike can be consumed in the subsequent acid copper-plating tank. Operating costs for the strike solution are 25% less than Watts nickel or copper pyrophosphate costs. High throw acid copper strike plating processes can be used over both electroless copper and nickel; however, a special conditioner is needed with an electroless nickel deposit. The conditioner prevents formation of "immersion patterns" created when electroless nickel is introduced directly into an acid copper strike. According to proponents of high throw acid copper strike plating, nickel strikes have the following disadvantages:

1. There is incomplete surface coverage due to poor throwing power.
2. Total operating cost is higher (e.g., nickel anode cost is three to four times more than copper).
3. Several additional steps are required prior to copper plating: (a) drag-out recovery stations, (b) multiple rinsing, and (c) sour rinses.
4. Effluent from rinsing the nickel strike must be treated to meet water pollution control requirements.
5. Metal deposited from a nickel strike does not contribute to the copper thickness specifications.

Copper pyrophosphates have better throwing power according to this evaluation, and copper does contribute to the total copper thickness specifications. However, high throw acid copper proponents cite the following disadvantages of conventional copper pyrophosphate:

1. Copper pyrophosphate solutions cost as much as or more than nickel solutions.
2. Rinsing is required between the strike and acid copper to protect electroplating solutions. Rinses must be waste-treated.
3. Waste treatment of copper pyrophosphate can be more difficult than waste treatment of nickel strike solutions because copper pyrophosphates are relatively strong complexing agents.

A comparison of conventional chemical copper with chemical nickel shows advantages and disadvantages for both metals. Copper solution is easier to activate for further electroplating; however, the copper can completely dissolve before electroplating begins if the copper coating is not thick enough. Inadequate etching can cause chemical copper to blister. Blistering can occur with inadequate activation of the metal. Nickel coatings can tend to delaminate under relatively small loads. Delamination occurs most often when no porosity develops during etching.

There are alternative and modified electroplating processes: for example, semibulk rather than straight-through bulk electroplating, direct use of trivalent chromium (Cr^{3+}) instead of using hexavalent chromium (Cr^{6+}) and reducing it to trivalent chromium, and microporous chrome plating.

Conventional straight-through systems use racks from which the plastic parts hang, while a semibulk system allows the parts to be closely packed onto metal trays for preplating. The parts can be inspected visually after the electroless copper plating process and placed on racks for electroplating. Semibulk was developed jointly by Shipley Company and Crown City Plating Company in the United States. According to the developers, semibulk plating equipment costs less and requires less energy; also, plated parts losses are lower and production can be increased up to 25%. Semibulk plating can produce the same quality plating performance with one fourth the nickel film thickness.

Trivalent chromium (Cr^{3+}) electroplating, developed by Harshaw Chemical Company, delivers microdiscontinuous deposits to improve corrosion protection in the finished plated part; and it eliminates burning, increases production, and eliminates waste by producing 85% less sludge. According to the developer, Cr^{3+} can deposit around sharp corners and recesses, and into holes with high aspect ratios, which would not be possible with Cr^{6+}. Other advantages with Cr^{3+} are (1) a partially plated rack can be removed from solution and parts can be observed visually; (2) burning at high current densities is eliminated, permitting the use of short, high current density strikes for improved covering power; (3) chromium metal deposits are more uniform; (4) air pollution is reduced because chromic acid is not produced and chromium hydroxide sludge is reduced by 85%, and (5) parts rejection is reduced, increasing productivity.

Microporous chrome plating, represented by the M&T Chemicals, Inc., "Micro-Dip" process, improves adhesion, corrosion resistance, and luster. A Micro-Dip solution contains a large number of small nonconductive particles that form more than 64,000 pores/in.^{2} (microporosity) over the plastic substrate. The Micro-Dip bath is followed by another (conductive) nickel plating bath. Udylite/Oxy Metal Industries produces microporous "Dur-Ni" and "Tri-Ni" solutions. Microporous plating is used for decorative applications.

Plating chemical removal systems are required to comply with the U.S. Water Pollution Control Act of 1972, to bring plating chemical effluents into surface waters down to zero content. Six methods to recover chrome and nickel plating solutions are (1) climbing film evaporation, (2) atmospheric evaporation, (3) submerged tube evaporation, (4) flash evaporation, (5) ion exchange, and (6) reverse osmosis. These systems remove the following four types of plating chemical losses: (1) drag-out, spills, fumes and mist, and dumping. Drag-out is the principal source of plating chemical loss, consuming 50% to 90% of every pound or gallon of plating chemical purchased by the plater. Recovery, therefore, refers primarily to capturing lost drag-out chemicals. Drag-out enters the first rinse tank where it is diluted. Dilution continues with each subsequent water rinse.

Electroless Plating

Electroless plating refers to plating processes that use chemical reductions without electrolytic anode–cathode reactions, therefore not requiring a galvanic reaction. The basic components of electroless plating are (1) copper or nickel concentrate such as nickel with phosphite solutions; (2) metal solubilizers and stabilizers such as heterocyclic organic sulfur compounds (e.g., 2-mercaptobenzothiazole); (3) reducing agents, especially formaldehyde in a basic (high pH) caustic soda (NaOH) solution; and (4) adjusters such as cyanide rate controllers.

Electroless plating is being used increasingly for producing conductive plastics for EMI/RFI shielding and circuit boards. The preplate process, represented by the Crown City Plating Company "Crownplate" process, consists of five steps: (1) etching, (2) neutralization, (3) catalysis, (4) acceleration, and (5) electroless copper or nickel deposition. The first four steps have similar functions to these steps in electrolytic plating. Electroless copper offers advantages over electroless nickel in wet corrosive environments, according to research at both General Motors Corporation Technical Center and Ford Motor Company. Electroless nickel loses its adhesion and tends to blister because of its preferential corrosion.

A principal reducing agent used in large-volume electroless copper systems is 3% formaldehyde + 10% methanol stabilizer in solution. The pH is maintained between 11 and 13, depending on the additive formulation. A 50% caustic soda solution is used to adjust the pH. Formaldehyde reduces copper ions to copper metal increasingly with higher pH. New electroless copper compounds use improved rate controllers and stabilizers, which permit higher processing temperatures without decomposition of the solution. The buildup of a solution that is characteristic of room temperature electroless plating is eliminated. Reduction of Cu^{2+} ions to metallic copper is a two-step process: divalent copper is first reduced to monovalent copper ions, which are further reduced to copper metal. Rate controllers such as cyanide, iodide, and non–sulfur-containing nitrogen heterocyclic compounds (bipyridyls and phenanthrolines) react with monovalent copper.

Copper deposition by autocatalytic reduction (electroless) costs more than electrolytically deposited copper, based on equivalent film thicknesses. Therefore, to save money electroless copper film thicknesses are kept below 1.0 μm. The optimum thickness for electroless copper films is approximately 0.75 μm. Electroless plating is a preplating step with electrolytic plating, but "stand-alone" electroless plating, omitting electrolytic plating, is used to produce conductive plastics for EMI/RFI shielding applications.

Electroless nickel is produced by reducing chemical nickel ions in solution to metallic nickel. Nickel ions are in solution with phosphites, hydrazines, and sometimes boron compounds. Alloys of nickel and phosphorus form with varying nickel-to-phosphorus ratios. High phosphorus content provides high strength and low stress, magnetic characteristics, and ductility.

Vacuum Metallizing

Evaporative Vacuum Metallizing Evaporative vacuum metallizing is primarily used for producing decorative plastics, but this metallizing process is being used more for EMI–RFI shielding. Evaporative metallizing uses aluminum or aluminum alloys for decorative applications and chromium–copper for EMI/RFI shielding applications, whereas sputtercoating uses chromium and chromium alloys. The metals used in evaporative vacuum metallizing go through two phase changes. The initial metal, such as chromium and copper, is evaporated in the vacuum chamber and condensed back to metal on the plastic substrate surface. The metal composition on the plastic, therefore, is different from that of the metal initially introduced into the vacuum chamber. Sputtercoated chromium composition, on the other hand, is unchanged from its initial state to its deposited state.

Both processes require lacquer basecoats and topcoats for decorative applications, but evaporative vacuum coating for EMI/RFI shielding applications does not require basecoats and topcoats, as explained further in this chapter.

Stokes Division, Pennwalt Corporation, is a principal producer of evaporative vacuum metallizing equipment. Sputtercoating equipment is made by Varian Associates, Inc., and Airco Temescal in the United States.

A vacuum is obtained with similar equipment in both evaporative metallizing and sputtercoating, using diffusion pumps and high-speed rotary pumps, with mechanical boosters. Evaporative vacuum chamber pressure is reduced to 1×10^{-6} atmospheric pressure. Firing pressure for aluminum is 5×10^{-4} Torr (0.5 μm). The vacuum chamber designs and construction, however, are completely different. Both processes can be either continuous or batch. Film thicknesses are considerably less than in electrolytic–electroless plating metal films. Sputtercoated chromium films are typically 3 μin. compared with 3000 μin. for typical electrolytic chrome plated films (600 Å = approximately 2.4 μin.).

Conventional evaporative vacuum-metallizing equipment must be modified to meet metallizing requirements for EMI/RFI shielding applications. To obtain a hard, shield-effective metallized coating, a three-layer chromium–copper–chromium coating must be applied. Conventional evaporative vacuum metallizing for decorative applications uses only one aluminum metallizing step. Conventional equipment is equipped with one bus-bar assembly. To apply two different metals, namely, chromium and copper for EMI/RFI shielding, the basic carriage of a conventional vacuum-metallizing chamber is modified by installing a second bus bar, center mounted as a single firing assembly.

The second modification of conventional equipment is a glo-discharge device for cleaning the plastic surface, developed by the Vacuum Equipment Department, Stokes Division, Pennwalt Corporation, Philadelphia, Pennsylvania. The plastic substrate surface is cleaned with an ionized gas plasma treatment prior to vacuum metallizing. Glo-discharge is the continuous bombardment of the plastic surface with ionized gas to remove contaminants such as residual mold carbon dioxide and even fingerprints. The ionic bombardment creates a highly reactive, clean surface that forms a strong bond with the chromium film. Glo-discharge is a plasma composed of a partially ionized gas, formed by applying a high negative voltage from an internal electrode to the chamber wall. The electromagnetic force (EMF) produced by the voltage ionizes some of the gas-phase molecules, forming the plasma. The positively charged gaseous ions are attracted to the negative electrode in the chamber. It is these gaseous ions that bombard the plastic surface, when the positively charged ions travel at high speed toward the negative electrode.

Substrate surfaces for decorative applications have a basecoat, therefore these surfaces do not require plasma cleaning. Metallizing for shielding applications does not require a basecoat and topcoat, because the chromium–copper–chromium film is 7500–15,000 Å thick, compared with 600 Å for decorative applications. An evaporative vacuum-metallized plastic substrate could consist of 500-Å thick chromium over the substrate, followed with 5000–10,000-Å copper deposit, and then 2000–5000-Å chromium deposit. Chromium forms a strong bond with most thermoplastics and with several thermosetting resins. Copper is highly conductive, providing effective EMI–RFI attenuation (shielding). The copper layer also provides conductivity for shielding in the electroless process. The final chromium deposition provides a durable, chemical resistant surface.

Sputtercoating

Sputtercoating is used primarily to metallize electronic component surfaces. Sputtercoating reduces plating process cost compared with electroplating, according to Varian Associates, not including initial capital investment for equipment. Sputtercoating requires approximately one third the electrical energy, and waste disposal costs are much less. Table 3.6 compares the steps for electroplating with those for sputtercoating.

A typical chrome sputtercoating layer can be 300 Å thick (minimum), not including the basecoat and topcoat. Below 300 Å, films become partially transparent. Up to 1000 Å, the metal film offers little abrasion resistance, and a urethane or acrylic lacquer topcoat is applied over the metal film. The upper limit to achieve uncracked chrome coating (film) is 2000–3000 Å. High deposition rates increase crack resistance, with a minimum rate of 50 Å/sec. Crack resistance is also affected by the distance between the disc targets in the chamber ceiling and the substrate. Longer distances reduce crack resistance because the kinetic energy of impacting chromium atoms is less and deposition rates are reduced.

The process became more efficient and cost effective with development of magnetically enhanced sputtercoating. Magnetically enhanced sputtercoating equipment has an electromagnetic field in the vacuum chamber, between the chrome disc targets on the ceiling of the chamber and the substrate on the chamber floor. The electromagnetic field is a "lens" that focuses chromium atoms sputtering from the chrome disc targets to the substrate. Without the electromagnetic lens focus, random sputtering of chromium atoms is not dense enough to provide efficient, cost-effective chrome coverage on a plastic substrate surface. Without the lens, only one chromium atom of every seven sputtered chromium atoms lands on the substrate, and the other six atoms coat the chamber wall.

Sputtercoating deposits both metals and nonmetals onto a substrate.

Table 3.6 Steps for Electroplating and Sputtercoating

Electroplating	Sputtercoating
Pre-etch bath	Anneal substrate
Acid bath	Apply basecoat
Neutralizer bath	Cure basecoat
Activator bath	Sputtercoat chrome: 2–3
Accelerator bath	Apply topcoat
Electroless copper or nickel	Cure topcoat
Dull nickel coating: 500 Min.	(Decorative applications)
Bright nickel coating: 500 Min.	
Chrome plating: 100 Min.	

The substrate can be metal or other materials, including plastics. Deposition is done under high vacuum in a chamber that contains positively charged argon gas molecules that are bombarded against the metal disc target, usually chromium or chromium alloys, causing the chromium atoms to sputter off the source target. The sputtering is analogous to hot oil sputtering randomly off the surface of a frying pan. The inert argon gas molecules are ionized by an electrical discharge. The argon gas is backfilled at a pressure of about 10^{-3} Torr. Although sputtercoating is done in a vacuum chamber, the process is completely different from evaporative vacuum metallizing. The metal is vaporized in evaporative vacuum metallizing, by either resistance heating or electron beam bombardment. The metal vapor molecules travel through the chamber and condense on the cooler substrate surface to form a metal coating. Sputtercoated metal atoms are not heated and melted, vaporized, and condensed. Instead, the metal (i.e., chrome) atoms are physically broken away from the disc target, atom by atom, by the force of the bombarding argon molecules.

The deposited metal is not the same as the original metal in evaporative metallizing because the original metal has vaporized and reformed on the substrate surface; while the original metal used in sputtercoating is the same metal that is deposited on the substrate surface, as noted earlier.

There are several ways to provide a magnetron sputtering source, which ionizes argon gas molecules to initiate the sputtering process. The magnetron sputtering source is an important factor in the efficiency of the process and the quality of the metallized product. The high kinetic energy (i.e., the high force of the impacting argon gas molecules on the disc target) causes chromium atoms to produce stronger mechanical adhesive bonding and denser (not thicker) coatings on the substrate surface, compared with evaporative vacuum metallizing. Varian Associates use an "S-Gun" magnetron sputtering source that consists of a water-cooled anode and a cathode from which electrons are emitted. The electrons collide with, and ionize, the argon gas molecules. Most of the electrons "miss" the argon molecules and are captured at the anode. Bombardment of excess electrons on the substrate surface would cause the surface to become hot, reducing both the bond strength and coverage of the metallized coating.

Disc targets are produced by Varian Associates, Airco Temescal (MH-557 standard target and HRC-885 large target inventory), and Emissive Products Group, GTE Sylvania (Ti-W targets). Homogeneous targets provide consistent sputtering density and uniform deposition of chromium (or alloys) on the plastic substrate.

Zinc Spray Coating

Conductive zinc arc spray is a principal method to achieve attenuation because it offers good conductivity, therefore effective shielding, over a wide electrical frequency range. Arc spray metallizing provides 60–90-dB attenuation, and more than 100-dB shielding effectiveness. It also provides a thick, durable, abrasion resistant metallic coating that is environmentally stable. The primary limitations of zinc art spraying are that it forms nonuniform films, it can create adhesion problems, and it generates zinc dust.

Zinc arc spraying uses two zinc wires that come into contact as the wires are fed through a spray gun. A dc arc, formed at the area of the gun where the ends of the two wires contact each other, melts the wires. An air jet atomizes and ejects the molten zinc onto the plastic substrate surface. The atomized zinc cools and solidifies on the plastic surface, forming a dense film 2–4 mil thick.

An alternative is a flame spray, but the arc spray system has a much higher efficiency. The higher efficiency of arc spray coating has two advantages: (1) it uses one ninth the melting energy of flame spraying, which requires fuel gas and oxygen, and (2) it reduces warpage and distortion of the plastic part.

Nickel Acrylic Lacquer Coating

These conductive lacquers are coated on plastics for EMI–RFI shielding applications. They are applied by conventional coating methods: spraying, brushing, dipping, and silk screening. The lacquers air-dry at room temperature in about 6 hours, and they can be force-dried at 150°F in 1 hour. Their service temperature range is −80–300°F. A typical nickel acrylic lacquer could provide attenuation of 30–60 dB over a 10–1000-Hz frequency range.

The metallic conductive lacquers are applied to plastic housings for electronic equipment such as computers, copiers, meter boxes, and so on, for isolating microelectronic devices that generate interference signals. Any unshielded ac device including home appliances such as microwave ovens, television sets, and vacuum cleaners can emit electromagnetic signals. Nickel acrylic lacquers are formulated by Master Bond, Inc.; Mereco Division, Metachem Resins Corporation; Red Spot Paint & Varnish Company; and Acheson Colloids Company in the United States.

Conductive copper lacquers, produced by Mereco for EMI/RFI shielding plastic housings for ac devices and appliances, form a relatively low cost, oxidation resistant, corrosion resistant coating. They show good adhesion to most solid and foam (cellular) plastics, including polycarbonate. The copper lacquers show no significant reduction in shielding effectiveness after humidity testing. They are thinned with an ethanol mixture and applied by spray gun.

Applications

EMI/RFI Shielding

The U.S. Federal Communications Commission (FCC) Docket 20780 requires that alternating current electrical devices manufactured in the United States after October 1, 1983, must be certified to comply with FCC radiation limits. Devices are divided into two categories: Class A for business, industrial, and commercial products and Class B for consumer products. The regulation applies to digital electronic devices that generate or use electrical frequencies between 10 kHz and 1000 MHz.

Metallized plating and coating are the primary methods to comply with FCC Docket 20780, by providing EMI shielding (RFI shielding is encompassed within the EMI shielding range). Conductive materials such as metals and carbon can be compounded into plastics to achieve shielding. Conductive compounds are described in another chapter. Shielding is achieved with conductive materials; therefore, metals are inherently shielding materials. Plastics are transparent to electromagnetic and radiofrequency (EM/RF) range radiation. With the increasingly widespread use of plastic housings for electrical and electronic devices, EM/RF range radiation is passing into and out of ac devices, causing malfunction of these devices, which include home appliances, computers, instrumentation, engine and machinery controls, and so on. Table 3.7 shows principal electrical and electronic products susceptible to EMI/RFI. The following list shows the principal electronic and electrical sources of EMI/RFI.

Sources of EMI/RFI

Business and home computers
Electronic calculators
TV and radio sets
Engine ignition systems
Induction heating units
Electric motors
Mobile communication transmitters
Paging systems
Remote control units
Relays and circuit breakers

Radar transmitters
CB transmitters
Static electricity
Lightning

The selection of EMI shielding processes depends on two factors: cost effectiveness and shielding effectiveness. These two criteria are further dependent on (1) unit production volume (e.g., several thousand units per year versus several 100,000 units per year), (2) product design (simple planar versus complex contoured design), and (3) available shield-coating capacity. The third factor is related to the increasing use of alternative shield-coating processes such as electroless plating, where market demand could be greater than available electroless processing plant capacity as of 1989.

FCC Docket 20780 requires that Class A and Class B devices meet specified limits for emitted radiation. The regulations are complex, with stricter limits for Class B products. Radiation limits are related electrical frequencies from 30 to 1000 MHz for radiated EMI, or 0.45 to 30 MHz for conducted EMI. FCC regulations do not specify methods to meet requirements, such as circuit designs and product designs. However, regulations apply to emissions from assembled products. Testing the shielding effectiveness of a plastic housing or enclosure can provide an indication of FCC compliance, but emission levels from the total enclosed device must be tested to ensure compliance with FCC Docket 20780. Therefore, redesigning is necessary for many products to comply with the FCC regulations. There are examples of assembled products with the

Table 3.7 Electrical and Electronic Products Affected by EMI/RFI

Class A	Class B
Data-processing equipment	Personal Computers
Industrial computers	Video equipment, including tv
Programmable controllers	Audio and high-fidelity equipment, including radios
Microprocessors	Appliances
Electronic typewriters	Cardiac pacemakers
Word processors	Digital Watches
Large calculators	Electronic games
Copiers	Pocket calculators
Telephones	Digital clocks
Remote control devices	CB receivers
Digital scales	
Navigation equipment	
Sensitive test instruments	

circuitry itself shielded effectively, so that the plastic housing does not need to shield the product from radiation emission. However, a radiation-transparent housing can still allow EM/RF emissions to enter the device and cause malfunction.

Three test methods have been developed for measuring radiation transmission through a conductive plastic product: (1) Stutz transmission test, (2) shielded box and (3) shielded room. The American Society for Testing and Materials (ASTM) has adopted methods 1 and 2. The Society of the Plastics Industry (SPI) and the American Society of Electroplated Plastics (ASEP) have committees devoted to EMI/RFI shielding testing and standards.

Shielding effectiveness is measured by the level of signal strength (decibels) blocked at varying electrical frequencies over a range from approximately 10 kHz to 1 GHz (gigahertz). The decibel scale is a logarithmic scale; therefore, 30 dB (minimum acceptable shielding effectiveness for many devices) is 10 times less than 60 dB. Shielding values are classified into five levels:

1. 0–10 dB, little or no shielding
2. 10–30 dB, minimal shielding
3. 30–60 dB, average shielding
4. 60–90 dB, above-average shielding, military uses
5. 90–120 dB, maximum to beyond state of the art

Translated to attenuation efficiency the following table relates decibel levels to attenuation (reduction of EM/RF transmission):

dB	Attenuation (%)
10	90
30	99.9
50	99.999
70	99.99999
90	99.9999999
110	99.999999999

A conductive plastic can provide electrostatic discharge (ESD) without providing EMI/RFI. The ESD and EMI/RFI capability is determined by surface resistivity (ohms/square area, e.g., ohms/sq cm, or simply ohms/sq). Surface resistivity values for ESD, RFI, and EMI shielding are as follows:

ESD, 50–100 ohms/sq
RFI shielding, >10 ohms/sq
EMI shielding, >1 ohm/sq

Standards for specific applications are being developed, for example, 5 ohm/sq for EMI shielding business machine housings.

Plastics are inherently insulators (i.e., nonconductive), with resistance values of 10^{14} to 10^{17} ohms. Conductive plastics can be classified as:

Antistatic, 10^{7}–10^{13} ohms
Static dissipating (ESD), 10^{3}–10^{6} ohms
EMI shielding, ~10^{2} ohms

Printed Circuit Boards

Copper metallized printed circuit boards provide better-quality products and cost less to make than the solder-plated boards they replace. Conventionally, special-grade epoxy thermosetting resins are used, but high-temperature engineering grade thermoplastics were introduced in the 1980s.

An example of circuit board grade epoxy resins is Ciba-Geigy Corporation Araldite 8011 flame retardant resins. They are used for rigid and multilayer printed circuit board laminates. The four epoxy resin grades in the series contain 19% to 23% bromine for flame retardancy. Each grade is soluble in different solvents, namely, acetone, methyl ethyl ketone (MEK), or methyl isobutyl ketone, with 75% to 80% (weight) solids. The characteristics of a good resin for circuit board bases are

1. Good copper peel strength
2. Good cold-punching properties
3. Good electrical properties
4. Good chemical resistance
5. Batch-to-batch property uniformity
6. Quality of varnish solution easily controllable in treater
7. Controlled flow and gel characteristics
8. Low squeeze-out and edge losses
9. Coefficient of thermal expansion and tensile strength (Table 3.8)

The resins are formed into prepregs by impregnating panel grade glass fiber cloth with a varnish solution. Properties of circuit board grade laminates, represented by fabric-reinforced Araldite 8011 prepared as a varnish with acetone, are shown in Table 3.9.

High-temperature thermoplastics such as polyaryl sulfone, polyether

Table 3.8 Tensile Strength and Coefficient of Expansion

Plastic substrate	Tensile strength, (psi × 10^3)	Coefficient of Thermal Expansion, (10^{-5} in./in./°F)
ABS EP-3510	5.5	3.9
ABS EPB-3570	6.4	3.6
ABS PG-298, PG-299	7.0	4.8
Caprez DPP	3.4	5.0
Cycoloy EHA	6.6	3.5
Noryl PN-235	7.0	3.5
Nylon 6 Capron CPN 1030	12.5	2.8
Nylon 6/6 Minlon IIC-40	13.0	3.1
	16.0	2.8
Polycarbonate Lexan 101	9.0	3.8
Polysulfone Udel P-6050	12.9	2.5
Derakane Vinyl ester	32.0	—
	25.0	
Polyester OCF E980/E573	25.5	—

sulfone, and polyphenylene sulfide are used for circuit boards for telecommunication, aerospace, and automotive applications. Flexible thermoplastic boards can be designed to meet unconventional configurations for interconnection systems. Multilayer boards can be designed into multiple circuit packages that occupy a space not much larger than a single circuit.

Copper metallized circuit boards replace solder-plated circuits because the copper circuit boards eliminate the undesirable characteristics of solder-plated circuits and reduce the number of processing steps. Undesirable characteristics eliminated are (1) "wrinkling," which occurs during wave soldering, (2) the solder mask tearing off the panel board, and (3) excess deposition caused by process solutions. Processing steps eliminated are (1) solder brightening, (2) cleaning prior to reflow, and (3) reflow. These three steps are eliminated from all three principal processes for fabricating copper circuit boards. Additional steps are eliminated for each of the three individual processes. The three processes are the following: (1) subtractive, (2) semiadditive, and (3) fully additive plating. Standard subtractive processes are most widely used. Semiadditive processes can be used in most facilities, but they require more steps than fully additive plating. The six principal semiadditive process steps are:

1. Electroless plate, 25–100 μin. thick
2. Image

3. Electroplate with copper to fill circuit thickness
4. Strip image
5. Etch
6. Solder mask

Several semiadditive methods combine additive and subtractive techniques. An entire panel is copper-plated by additive methods and subsequently processed as a conventionally produced flash-plated board. An advantage of this method is that a plater can prepare base boards for circuit boards by adding one or two electroless solution tanks. In the subtractive method for through-hole plating, the electroless copper has only one function: to make the plastic substrate conductive, preparing the surface for electrolytic plating.

Fully additive processes are probably the most efficient but least popular techniques. The entire circuit pattern and plated through-holes are created by electroless copper plating. No electrolytic plating is used, as in EMI-shielding metallizing with only electroless plating. To achieve a thickness of 1–1.5 mils the immersion time in electroless copper plating is typically 12 to 20 hours. The advantages and disadvantages of fully additive plating are as follows:

Advantages
Virtually no subsequent steps required
Excellent line definition and circuit density
Minimal waste treatment of solutions
No problems related to electrolytic plating
Uniform plating of blind holes
Lower labor costs: no deburring, sanding, or reracking
Less rejects
Repairable: boards can be stripped and replated

Disadvantages
Long dwell time in the electroless copper bath
Tight control required for electroless plating solutions
Dependence on quality of the plating resist

Additive plating consists of two types of surface preparation prior to electroless plating: (1) coating the plastic substrate surface or (2) chemically pretreating the plastic substrate, as in conventional plating-on-plastics pretreatment of plastic substrate surfaces. Precoating permits the use of the same coating material over a wide variety of substrates but requires higher initial capital investment. Chemical pretreatment can

Table 3.9 Circuit Board Grade Epoxy Laminate Properties

Fabric—finish	HG-28—Volan A
Plies	9
Cure	30 min/350°F (177°C)/500 psi
Resin content	30%
Flexural strength, psi (kg/cm^2)	85,000 (6000)
Elastic modulus, psi (kg/cm^2)	2.53×10^6 (1.78×10^4)
Water absorption, % 24 hr/73°F	0.09
Peel strength, lb/in. of width after solder (2 oz Cu)	12
Flammability	
Seconds to extinguish	0.0
Extent of burning (i.e., VO)	0.0

require different chemicals and pretreatment cycles, but it is easier and lower cost.

The electroless copper solution is the most essential phase in additive processes for quality-performance circuit boards. The copper solution should have the following attributes:

1. Rapid plating rate
2. Excellent stability
3. Excellent reliability
4. Yield of a deposit with specific, uniform physical and mechanical properties

The selection of an electroless process is related to high aspect ratio (length–diameter ratio) holes and fine line circuitry. The basic changes in a copper electroless plating bath for high aspect hole–fine line circuitry, compared with a typical bath, is an increase in the copper sulfate–sulfuric acid ratio. The bath also contains a brightener and a small amount of leveling agent to achieve a stress-free deposit.

Chapter 4

Conductive Plastics

William M. Wright and George W. Woodham

The plastics industry is becoming very sophisticated, and growth is being driven by new product innovations and technical advancements. The estimated volume of plastics for 1988 was 51.5 billion lb. That will be a 77% increase since 1984. There is no question about the future of plastics!

Entirely new thermoplastic resins have recently been introduced by the industry. Blending, alloying, and reactive processing of existing resins are increasing, as is sophisticated compounding with reinforcements and modifiers as methods of improving properties and performances of these resins.

Much of the growth is attributed to the displacement of metal parts. This is true in the electronics industry, with the advent of miniaturization, and in the transportation industry, with the development of low-cost, low-weight, high-strength plastics that can withstand both processing and end use environments. The appliance, medical, business machine, packaging, and aerospace industries are also finding increased demands for plastics.

Thermoplastic compounds are molding materials composed of one or more base resins and other constituents, including stabilizers, reinforcements, wear-resistant additives, conductive additives, flame-retardant additives, and coloring. These "other" constituents modify the base resin's physical, mechanical, electrical, and/or thermal properties to address specifically an application's requirements. This "tailoring" ability of thermoplastic compounders is another contributing factor to the growth rate.

Most plastics are inherently excellent electrical insulators. The thermosetting phenol formaldehyde resins ("Bakelite" or "phenolic," as they were or are known) that evolved in the early 1900s and that are still in wide use today found rapid and successful acceptance as electrical insulation materials for circuit breakers and switches. Table 4.1 depicts where plastics—in this case thermoplastics—fit in a hierarchy of conductivity, which is the inverse of resistivity. (The use of the word *plastics* in this chapter refers strictly to thermoplastics.)

If thermoplastics are such outstanding insulators inherently and if good or excellent conductors such as aluminum and copper are already available, fairly inexpensive, and well understood, why modify thermoplastics to make them conductive? There are several reasons and they are chiefly a derivative of the cost–performance relationship of plastics to metals. Conductive plastics can be more cost effective than metals in some applications requiring electrical conductivity. The economics of the injection molding process is one reason for this advantage. Another is the design capability of combining several metal fabrication steps into a single molding step with plastics.

In addition, conductive plastics offer significant performance and design advantages over metals. For instance, the superior impact resistance and noncorrosiveness of plastics versus metals usually holds true for conductive plastics. Also, as will be discussed in more detail later, plastics can be formulated to fit varying levels of conductivity. In one major end-use area of conductive plastics, this "tailoring" of conductivity is very important to the performance of the conductive plastic part.

The fact that conductive plastics can be more cost effective and perform better than metals is meaningless if there is no real market potential for these benefits. As is often the case with plastics in general, tangible applications have driven the development work in conductive plastics. Early needs that have become extremely important to the electrical–electronics industry are electrostatic discharge (ESD) protection and electromagnetic–radio frequency interference (EMI/RFI) protection.

Static electricity can be simply stated as an excess or deficiency of electrons on a surface. It occurs when two nonconductive bodies rub or slide together or separate from one another. It is often referred to as the triboelectric effect. One of the materials will result in a positive electrical charge on its surface and the other, a negative charge. A number of factors

Table 4.1 Electrical conductivity of selected metals vs. thermoplastics

	Conductivity (ohm^{-1}–cm^{-1}) at 20°C
Silver	6.3×10^5
Copper	5.9×10^5
Aluminum	3.6×10^5
Nickel	1.5×10^5
Carbon (amorphous)	300
Conductive thermoplastic compounds	10^{-11}–50
Thermoplastics	10^{-17}–10^{-14}

influence the polarity and size of the charge—cleanliness, pressure of contact, surface area, and speed of rubbing or separating.

A second source of static electricity is an electrostatic field, formed by charged bodies when they are near each other or near noncharged bodies. This field will induce a charge on a nonconductive object it is near.

Prime generators of electrostatic voltages are plastics, synthetic fibers, fiberglass, and human bodies. Under proper conditions, these nonconductors can carry a charge as high as 35 kV. (See Table 4.2.)

This electrostatic voltage can be discharged when the nonconductive body carrying the charge comes in contact with another body at a sufficiently different potential. The discharge may be in the form of an arc or spark. Sudden, random discharges can be damaging to microelectronic parts or can cause an explosion in an environment containing flammable gases or liquids or in the manufacture of explosives.

With recent advances in integrated circuit technology, thousands of transistors are crowded onto a single chip by microminiaturization. To increase speed and reduce power consumption, junctions and connecting lines are finer and closer together. The resulting ICs are more sensitive and can be easily damaged by static discharge, and at very low voltages.

Modification of thermoplastics by the addition of a conduction material to the resin matrix results in a product or compound having conductive properties that can be used for protection against electrostatic discharge. The advantages of using a thermoplastic apply to the conductive thermoplastic compounds. Light weight, cost effectiveness, appearance, and convenient processing and finishing are among these advantages.

EMI/RFI Phenomena

Federal Communications Commission (FCC) regulation docket 20780, Part 15, issued in October of 1983, sets the electromagnetic regulation limits for radiation generated by computing devices. (See Table 4.3.)

Table 4.2 Typical Electrostatic Voltages (DOD-HDBK 263)

Means of Static Generation	10% to 20% Relative Humidity	65% to 90% Relative Humidity
Walking across carpet	35,000	1,500
Walking over vinyl floor	12,000	250
Worker at bench	6,000	100
Vinyl envelopes for work instructions	7,000	600
Common poly bag picked up from bench	20,000	1,200
Work chair padded with polyurethane foam	18,000	1,500

Table 4.3 RFI Radiation Limits FCC Regulation (47 CFR Part 2)

Class A Computing Device (commercial, industrial, business)		
Frequency (MHz)	Distance (meters)	Field Strength (micro V/M)
30–88	30	30
88–216	30	50
216–1000	30	70
Class B Computing Device (home or residential)		
Frequency (MHz)	Distance (meters)	Field Strength (micro V/M)
30–88	3	100
88–216	3	150
216–1000	3	200

Class A computing devices are defined as those marketed for use in business.

Class B computing devices are defined as those marketed for use in residential environments. Examples are home computers, electronic games, and electrical organs.

The definition of a computing device is any electronic device that generates or uses timing signals at a clock rate greater than 10 kHz and that uses digital techniques.

Thermoplastics are commonly used to make electronic enclosures and cabinetry. Because of their inherently high electrical resistivity, they are transparent to electromagnetic radiation and hence provide no protection against electromagnetic emissions of electronic devices. To take advantage of the attributes of thermoplastics—cost effectiveness, light weight, appearance, ease of processing—development efforts in recent years have resulted in commercially available conductive thermoplastic compounds that are acceptable as shields against EMI.

In addition to compounding conductive additives to impart EMI shielding characteristics to thermoplastics, conductive coatings have been applied to the surface of molded parts. A number of commercial methods of this type are available:

Conductive paints—silver, copper, nickel, and graphite
Vacuum metallizing
Arc spraying
Spray plating
Cathode sputtering
Electroless plating–electroplating
Pressure-sensitive foils

These all work, and some are more efficient than others. However, a number of disadvantages exist with these coating methods:

Adhesion to the substrate may be poor.
Surface scratches could cause EMI loss.
Geometry of the part may be difficult to cover uniformly.
There may be difficulty in coating around sink marks (a common molding problem).
Resins may be sensitive to any solvents used.
The thermal coefficient of expansion for the metal coating much higher than plastic could cause separation in cyclic thermal situations.
The method may require significant capital investment or special equipment.
Skilled operators may be required.
The coating thickness may be insufficient for good conductivity.
The process may require multiple steps.
The method may be limited to a few resin systems.

The use of conductive thermoplastics can eliminate the secondary finishing steps and any special equipment and skilled labor that are generally required to apply a continuous conductive layer onto the surface of a molded part.

The section of this chapter entitled "Conductive Carbon Powder" will help the user of conductive polymers select the material that best suits his application by identifying the classes of conductive compounds and evaluating them by polymer types, conductive fillers, and conductive applications.

"Carbon Fiber Additives" discusses test methods and the importance of test standardization. Varying methods now in use have resulted in resistivity values that do not correlate with each other.

Concerns involved with the processing of conductive thermoplastics in the injection molding industry are identified in the section entitled "Aluminum Flake."

Classes of Conductive Compounds

The U.S. Department of Defense's Handbook 263 (DOD-HDISK-263) describes three categories of plastic composites for use in ESD protection. They are antistatic, static dissipative, and conductive. These reference categories, although not universally accepted, are used by most leading companies involved in ESD protection.

The DOD usage of the term *conductive* is more precise and limited than the general use of the word *conductive* up to this point. In its most familiar form the word refers to any plastic that's been modified to be more conductive than it is in its natural form. When referring to actual range of conductivity, however, the reader needs to understand that the term *conductive* is narrowly defined.

Antistatic composites are defined as having a surface resistivity of greater than 10^9 and less than 10^{14} ohms/square. Commercial grades of compounds that fall into this range of conductivity are available in several resin systems. Although commercial grades are usually made by compounding solid conductive additives into a resin matrix, antistatic composites contain a liquid organic additive, such as a fatty acid amine derivative. Because the liquid antistatic materials are not permanently antistatic, the liquid agent tends to migrate out of the part over time.

Static dissipative composites, by definition, have a surface resistivity of greater than 10^5 and less than 10^9 ohms/square. Since static dissipative composites are more conductive than antistatic composites, they will dissipate electric potential more quickly. Also, because of their makeup, they will more readily conduct throughout the volume of the part. Thus, static dissipative composites can be used for more rapid bleed-off of electric potential. Commercial grades of static dissipative composites are available in a wide range of resin types, containing a number of different conductive additives. The static dissipative compounds are considered to be permanent, not changing their electrical characteristics with time, based on the fact that they are formed by blending a solid conductive material with the resin matrix.

Conductive composites for ESD protection have a surface resistivity of less than 10^5 ohms/square and are the most highly conductive category. These materials are normally used in applications requiring electrostatic shields, such as a Faraday cage effect. A molded part made from a conductive composite will rapidly dissipate electric potential and will shield electrostatic-sensitive components from electrostatic fields. However, electrostatic-sensitive components would have to be insulated from direct contact with a part molded from a conductive composite. This can be accomplished by using an antistatic conductive composite as the barrier material.

Compounds According to Conductive Additives

Suppliers of conductive thermoplastic products use a number of conductive additives. The following are examples of these conductive additives:

Reinforced Type	Additive Type
PAN carbon fiber	Carbon black powder
Pitch carbon fiber	Aluminum flakes
Nickel-coated carbon fiber	Metal powders
Stainless steel fibers	Organic-antistatic
Aluminum fibers	Metal-coated glass beads
Metallized glass fibers	Metal-coated mica

The reinforcement type of conductive additive not only provides conductivity when compounded into thermoplastic resin, but also enhances the strength properties of the base resin. Filler grades, on the other hand, generally reduce the properties of the base resin (although increased stiffness can occur).

The combination of the many conductive additives, either filler or fiber type, compounded into the different families of thermoplastic resins results in an endless number of commercially available conductive compounds. In addition, the ability to combine reinforcements with the conductive additive or to combine two or more conductive additives with a particular resin adds to the list of available products. Having many raw materials to work with, suppliers can "tailor-make" cost-effective products for specific applications.

Organic Antistat Additives A number of antistat compounds are commercially available based on an organic additive. The antistat agent functions by exuding, migrating, or "bleeding" to the surface of the molded plastic part. As a result, they can be removed by wiping or washing, but the antistatic properties will be regenerated as the additive migrates to the surface. Once on the surface, water is absorbed from the atmosphere by the antistat composites, which work best at high relative humidities. At relative humidities of 15% or less they generally do not function well. Because of their migratory and regenerative characteristics, they are classified as nonpermanent antistatic materials. These organic antistats must be processed at low temperatures. Because of that, only polypropylene-, polyethylene-, and polystyrene-based resin products are commercially available. Considerable development work is now under way to develop systems useful in higher-temperature resins.

Products containing antistat organic additives are characterized by transparency; colorability, if needed; and good retention of the polymer's original properties. Several examples of commercially available products follow:

Property	ASTM Test Method	Resin		
		PP	PS	HDPE
Specific gravity	D-792	0.905	1.06	0.96
Molding shrinkage				
in./in., 1/8 in. section	D-955	0.015	0.005	0.020
Impact strength, IZOD ft. lb./in.				
notched 1/8 in.	D-256	0.6	0.4	0.8
unnotched 1/8 in.		No break	2.0	No break
Tensile strength, psi	D-638	5100	6000	4200
Tensile elongation, %	D-638	150	1.5	250
Flexural strength, psi	D-790	7000	10,000	4000
Flexural modulus, psi $\times 10^6$	D-790	0.18	0.45	0.2
Surface resistivity, ohm/sq	D-257	10^9–10^{13}	10^9–10^{13}	10^9–10^{13}
Static decay				
Mil B-81705B, 2.0 sec	FTMS-101B Method 4046	pass	pass	pass
NFPA code 56A. 0.5 sec	FTMS-101B Method 4046	pass	pass	pass
Deflection temp., °F,				
at 264 psi	D-648	130	175	120
at 66 psi		225	205	160
Flammability, in./min	D-635	B	B	B
Flammability	UL94	HB	HB	HB

Products in the static dissipative or conductive range are not available with antistat additives. The need for high concentrations of the organic additive results in poor compatibility with the resin matrix.

Conductive Carbon Powder Commercial products containing conductive carbon powder make up the greater portion of all the conductive resins available. Economics plays an important role since this additive is the most cost effective. In addition, by varying the concentration of the carbon powder, products that fall into the three different groupings—antistatic, static dissipative, and conductive—are made available. Virtually all thermoplastic resins can be compounded with carbon powder. This provides the end user with a wide spectrum of product properties to select from.

Although some grades of carbon powder contain excessive sulfur that could corrode electrical contacts, grades are available with low concentrations of free sulfur. For very critical applications, extremely low-sulfur carbon powder–filled resin systems can be used.

Because carbon black is indeed black in color, no colored products are possible. As a general rule, physical strength properties of the base resin are reduced when carbon powder is added. The degree of loss is proportional to the concentration; the higher the loading, the lower the properties. Suppliers of these compounds overcome this problem to some degree by adding impact modifiers or using special grades of resin with enhanced properties.

Carbon powder–filled products tend to slough carbon powder off from the surface of a molded part; as a result, they cannot be used in clean room environments or other sensitive areas where contamination cannot be tolerated. Although products that minimize this problem are commercially available, the risk of contamination still exists, precluding their use in strict clean room environments.

Table 4.4 lists examples of the types of commercial compounds that are available containing conductive carbon powder.

Table 4.4 Conductive carbon black–filled products

		Resin		
Property	ASTM Test Method	PP	Impact Nylon	HDPE
Specific gravity	D-792	0.97–1.01	1.15–1.18	1.02–1.05
Molding shrinkage				
in/in, 1/8 in. section	D-955	0.017	0.016	0.022
Impact strength, IZOD ft. lb./in.				
notched 1/8 in.	D-256	8	1.5	1.2
unnotched 1/8 in.		No break	15	No break
Tensile strength, psi	D-638	3000	8000	3600
Tensile elongation, %	D-638	20	4–6	10+
Flexural strength, psi	D-790	3900	12,000	4300
Flexural modulus,				
psi $\times 10^6$	D-790	0.15	0.40	
Surface resistivity, ohm/sq	D-257	$<10^5$	$<10^5$	$<10^5$
Static decay				
Mil B-81705B, 2.0 sec	FTMS-101B Method 4046	Pass	Pass	Pass
NFPA code 56A, 0.5 sec	FMS-101B Method 4046	Pass	Pass	Pass
Deflection temp., °F				
at 264 psi	D-648	160	160	120
at 66 psi		260	420	160
Flammability, in./min	D-635	B	B	B
Flammability	UL94	HB	HB	HB

Conductive carbon black products are not acceptable as EMI/RFI shielding materials, because of the very high loadings of carbon black required to obtain the conductivity needed. Very poor physical properties result from the high loadings.

Carbon Fiber Additives Polyacrylonitrile (PAN) carbon fibers act as both a conductive additive and a reinforcement when compounded into thermoplastic resin. Products with varying levels of PAN carbon fiber are available in the three classes of conductivity. However, they are predominantly used to prepare products with surface resistivities ranging from 10^1 to 10^8 ohms/square. The availability of carbon fiber with different sizing systems allows the optimization of reinforcement in a variety of resin systems while at the same time achieving the desired conductivity for the product. The different sizing systems aid in chemically bonding the fibers to a particular resin. The following table illustrates the reinforcement characteristic by comparing properties of carbon fiber and carbon black compounds of an impact-modified 6-6 nylon.

		Conductive Additive	
Property	ASTM Test Method	Carbon Fiber	Carbon Black
Specific gravity	D-792	1.22	1.15–1.18
Mold shrinkage			
in./in., 1/8 in. section	D-955	0.001	0.016
Impact strength, IZOD ft. lb./in.			
notched 1/8 in.	D-256	3.0	1.5
unnotched 1/8 in.		14	15
Tensile strength, psi	D-638	25,000	8000
Tensile elongation, %	D-638	2–3	4–6
Flexural strength, psi	D-790	38,000	12,000
Flexural modulus, psi × 10^6	D-790	2.0	0.40
Surface resistivity, ohms/sq	D-257	75	< 10
Static decay			
Mil B-81705B, 2.0 sec	FTMS-101B Method 4046	pass	pass
NFPA code 56A, 0.5 sec	FTMS-101B Method 4046	pass	pass
Deflection temp., °F			
at 264 psi	D-648	395	160
at 66psi		430	420
Flammability, in./min	D-635	B	B
Flammability	UL94	HB	HB

EMI/RFI shielding materials with volume resistivities of 1 ohm/cm or less can be made with PAN carbon fiber. These materials have shielding effectiveness values greater than 30 dB. As previously stated, physical properties are greatly improved.

Metal-coated carbon fiber, and in particular nickel-coated fiber (NCG), provides another approach to make thermoplastics conductive. The NCG fibers are said to be 50 times more conductive than the PAN carbon fibers. Less highly conductive NCG fibers are required to achieve the same conductivity. In addition, these materials can be colored. Composites with very high conductivity, less than 1 ohm/cm volume resistivity, can be produced. These products are excellent candidates for EMI/RFI shielding applications. A typical commercially available line of products would contain concentrations of NCG fiber ranging from 10 to 40 wt % of the fiber in a thermoplastic resin system. Properties for such a product line based on polycarbonate are:

	ASTM Test Method		
Nickel-coated graphite fiber, %		15	20
Specific gravity	D-792	1.31	1.35
Mold shrinkage in./in. 1/8 in. section	D-955	0.002	0.001
Impact strength, IZOD ft. lb./in.			
notched 1/8 in.	D-256	1	1.1
unnotched 1/8 in.		6	6
Tensile strength, psi	D-638	12,000	14,000
Tensile elongation, %	D-638	3	2
Flexural strength, psi	D-790	16,000	18,000
Flexural modulus, psi $\times 10^6$	D-790	0.9	1.1
Volume resistivity, ohm-cm	D-257	0.5	0.1
Deflection temp., °F			
at 264 psi	D-648	290	290
at 66 psi		300	300
Flammability, in./min	D-635	SE	SE
Flammability	UL94	V1	V1
Shielding effectivenes (SE) dB at 1000 MHz	ES-7-83	30–40	40–60

This property data demonstrates the choices the design engineer has in selecting the best product for an application. A similar line of products can be made available from other resin types, each having properties based on the resin used. As an example, for applications requiring a material with high-temperature properties, NCG fiber–reinforced polyphenylene sulfide (PPS) can be considered as a possible candidate. Typical properties for this type of conductive product are as follows.

	ASTM Test Method		
Nickel-coated graphite fiber, %	D-792	15	20
Specific gravity	D-955	1.45	1.50
Mold shrinkage,			
in./in., 1/8 in. section	D-955	0.003	0.002
Impact strength, IZOD ft. lb./in.			
notched 1/8 in.	D-256	0.8	1.0
unnotched 1/8 in.		2.0	3.0
Tensile strength, psi	D-638	11,000	14,000
Tensile elongation, %	D-638	0.7	0.7
Flexural strength, psi	D-790	16,000	20,000
Flexural modulus, psi $\times 10^6$	D-790	0.3	1.6
Volume resistivity, ohm-cm	D-257	0.5	0.1
Deflection temp., °F			
at 264 psi	D-648	470	500
at 66 psi		500	500+
Flammability, in./min	D-635	SE	SE
Flammability	UL94	VO	VO
Shielding effectiveness (SE)			
dB at 1000 Mz	ES-7-83	30–40	40–60

Aluminum Flake Aluminum flake–filled thermoplastics are the most cost-effective system for EMI shielding applications. Their low cost compared with the other conductive additives and their ability to provide SE values of 40 or higher make them excellent candidates.

To achieve an SE of 40, it is necessary to utilize compounds containing 40 wt % or more of the aluminum flake. The high loading leads to several disadvantages—high specific gravity, poor surface finish (mottled surface), and poor mechanical properties—with the aluminum flake acting as an additive and not as a reinforcement.

Although these properties may prohibit the use of aluminum flake as the conductive additive for some applications, nonstructural applications where mechanical strength is not a requirement can take advantage of this low-cost conductive additive system. In addition, paint systems are available to cover the poor surface finish.

Aluminum flake–filled composites offer an added advantage of being highly thermally conductive and have found major uses in applications that use this property. A comparison of thermal conductivity for several materials shows how effective these materials are in dissipating heat.

Material	K (BTU/hr ft^2°F/in.)
Aluminum	1536
Aluminum flake–filled ABS	69
Carbon fiber–reinforced ABS	3.8
Glass fiber–reinforced ABS	1.6
ABS resin	1.3

Lab tests show that the 40 wt % aluminum flake–filled composites are 87% as effective as aluminum itself in transferring heat, making them excellent candidates for applications where both EMI/RFI shielding and thermal conductivity are required.

Aluminum flake–filled polypropylene and polyester composites can be used in applications to dissipate heat away from LED and LCD digital readouts that are affected by heat. When compounded into engineering plastics having high heat properties and good solvent resistance, they find use in under-the-hood applications, such as pulleys, bearings, and end brackets for motors.

Stainless Steel Fibers A variety of resin types of conductive products are available containing stainless steel fibers as the conductive additive. These compounds have several advantages compared with the others. All are related to the low concentrations of stainless steel needed to achieve the desired conductivity. Specific gravity is low, structural strength and impact properties are relatively unchanged from unmodified resin, and the compounds can be easily colored even in light shades. Processing characteristics are essentially the same as for the base resin. Also, a major advantage is that the mold shrinkage is changed very little from the base resin, which can allow the use of existing molds.

Because of the low concentrations of conductive additive, surface resistivity measurements may indicate that the material is nonconductive when in reality it is inherently conductive throughout its volume. Processing stainless steel composites requires some degree of caution to prevent fiber breakage, although the ductibility of the fibers minimizes this problem. Proper dispersion of the fibers throughout the molded part with a minimum of fiber breakage is the key to successfully processing a part with the desired shielding effectiveness.

Although these types of products have excellent static charge dissipation, they are used predominantly in applications requiring EMI/RFI shielding. Physical properties for a stainless steel–filled ABS product follow and are compared with the properties of the ABS resin.

	ASTM Test Method		
SS fiber concentration, %		0	7.5
Specific gravity	D-792	1.04	1.13
Mold shrinkage, in./in., 1/8 in. section	D-955	0.004	0.005
Impact strength, IZOD ft. lb./in.			
notched 1/8 in.	D-256	3.0	1.4
unnotched 1/8 in.		25	20
Tensile strength, psi	D-638	6000	6500
Tensile elongation, %	D-638	8	6
Flexural strength, psi	D-790	10,000	9500
Flexural modulus, psi $\times 10^6$	D-790	0.3	0.36
Volume resistivity, ohm-cm	D-257	10^{16}	10^0
Surface resistivity, ohm/sq	D-257	10^{16}	10^5
Shielding effectiveness, dB	—	—	35–40

Metal-Coated Fillers In addition to metal-coated fibers such as the nickel-coated graphite fiber, metal-coated fillers are used as conductive additives. One such filler is nickel-coated mica. Products containing 40 wt % for ESD applications and 50 wt % for EMI shielding applications are available. Different ESD requirements can be met by varying the filler loadings. These composites are free from the sloughing problem experienced with carbon black. In addition, they can be pigmented with colors. The surface finish is improved over that of aluminum flake. Being a filler and not a reinforcement, physical properties are lower than the base resin.

% Nickel-Coated Mica	10%	20%	30%	40%	50%	60%
Tensile strength, psi	4230	4376	4763	5123	5336	5310
Flexural strength, psi	5492	6072	6564	6890	7732	8433
Flexural modulus, psi $\times 10^5$	3.2	5.0	7.3	9.1	1.16	1.29
Impact strength						
notched ft-lb/in.	—	—	0.82	0.76	0.65	0.66
unnotched ft-lb/in.	—	—	5.6	3.2	2.9	2.3
HDT_1 °C 264 psi	74	84	92	105	116	130
Volume resistivity (ohm-cm)	1×10^9	4×10^7	9.8×10^0	8.4×10^{-1}	1.6×10^{-1}	1.2×10^{-1}
Surface resistivity (ohm/sq)	3.3×10^{10}	2.3×10^2	1.6×10^0	9×10^{-2}	4.2×10^{-2}	3.8×10^{-2}
Shielding effectiveness (DB)						
100 MHz	3.6	18.0	50.4	62.8	67.6	71.4
300 MHz	3.3	15.8	42.6	53.8	58.7	61.1
500 MHz	3.7	14.1	37.9	48.9	54.0	55.6
1000 MHz	3.8	13.9	33.5	42.6	55.6	50.7

Tailor-Made Composites A combination of conductive additives and other modifiers can be used to produce a variety of hybrid thermoplastic composites. The availability of these "tailor-made" products enables the end user to choose the most cost-effective system that meets the requirements of an application.

Table 4.5 gives an example of static dissipative polypropylene composites containing different reinforcements and fillers in combination with carbon black. Fiberglass greatly increases tensile strength, flexural strength, and flexural modulus while at the same time reducing shrinkage. Flexural modulus is enhanced by the addition of glass beads or mineral fillers and shrinkage is reduced, but not to the same degree as with fiberglass. Specific gravity is increased with all fillers and reinforcement. Fiberglass may cause a molded polypropylene product to warp, depending on the mold design. Fillers having very low aspect ratios will produce more dimensionally stable parts.

The properties for the products shown in Table 4.5 can be varied by

Changing the concentration of the reinforcement or filler
Changing the type of reinforcement or filler

Table 4.5 Carbon powder–filled propylene products containing other fillers or reinforcements

Property	ASTM Test Method	Filler or Reinforcement: Fiberglass 20%	Filler or Reinforcement: Mineral 20%
Specific gravity	D-792	1.09–1.12	1.09–1.12
Mold shrinkage			
in./in., 1/8 in. section	D-955	0.004	0.012
Impact strength, IZOD ft. lb./in.			
notched 1/8 in.	D-256	2.5	5
unnotched 1/8 in.		6	20
Tensile strength, psi	D-638	5000	2000
Tensile elongation, %	D-638	3–5	10+
Flexural strength, psi	D-790	7000	3000
Flexural modulus, psi × 10^6	D-790	0.42	0.17
Surface resistivity, ohms/sq	D-257	$<10^5$	$<10^5$
Static decay			
Mil B-81705B, 2.0 sec	FTMS-101B Method 4046	Pass	Pass
NFPA code 56A, 0.5 sec	FTMS-101B Method 4046	Pass	Pass
Deflection temp., °F			
at 264 psi	D-648	190	160
at 66 psi	—	220	210
Flammability, in./min	D-635	B	B
Flammability	UL94	HB	HB

Changing the grade of polypropylene
Adding impact modifiers

Hybrid formulations of conductive additives with fillers or reinforcements are made available to achieve specific mold shrinkage, greater reinforcement characteristics, and improved lubricity—all formulated for the desired surface resistivity.

Table 4.6 shows several flame-retardant (FR) static dissipative products. These products combine the conductive additive with special flame retardants to achieve the conductivity and flame-retardant properties desired. In addition, reinforcements are combined with the additives in two of the products to enhance properties. Property comparison between the modified 6-6 nylon– and propylene-based resin products, both unreinforced, show the enhanced properties achieved with the 6-6 nylon resin. High flexural modulus FR composites are achieved by reinforcing with fiberglass, attested to by the property data shown for the conductive PBT (polyester).

The FR conductive products shown are only a few selected FR products from the many that can be made available based on combining the different conductive additives, reinforcements, and fillers with the flame-retardant additives. In addition, resins such as PPS, PES, PEI, and PEEK are inherently flame retardant. Conductive products based on these systems will be flame retardant, having a UL subject 94 V-0 rating.

The following table lists conductive polyphenylene sulfide products having very high flexural modulus values.

Resin	Carbon Fiber Type	Flexural Modulus (psi × 10^6)	Volume Resistivity (ohm/cm)
PPS	None	0.550	10^{14}
PPS	PAN	4.6	10^1
PPS	PAN (high modulus)	6.0	10^1

These composites have modulus values eight to 10 times greater than the base resin. Specialty products are made available with a range of modulus values by varying the PAN carbon fiber content in the product. Using a resin such as PPS, a flame-retardant product (UL subject 94 V-0) is created that can also be used in high-temperature applications.

The high-modulus PPS-based products are an example of what can be made available using other resin types. Resulting modulus values will depend on the resin type and the degree of reinforcement achieved with the PAN carbon fiber.

Table 4.6 Flame-retardant static dissipative thermoplastic resin composites

Property	Units	Test Method	Resin			
			PP	PC	PBT	Modified 6-6 Nylon
Conductive additive	—	—	Carbon black	Carbon black	Carbon black	Carbon black
Reinforcement–filler	—	—	None	Fiberglass	Fiberglass	None
Specific gravity	—	—	1.12	1.31	1.66	1.20
Mold shrinkage	in./in.	D-955	0.016	0.003	0.002	0.014
Tensile strength	psi	D-638	2500	7000	15,000	7000
Flexural strength	psi	D-790	4000	12,000	23,000	11,000
Flexural modulus	psi $\times 10^6$	D-790	0.180	0.550	1.40	0.400
IZOD impact	ft-lb/in. notch	D-256	2.5	1.3	1.3	1.2
Surface resistivity	ohms/sq	D-257	10^5–10^8	10^5–10^8	10^5–10^8	10^5–10^8
Volume resistivity	ohms-cm	D-257	$<10^5$	$<10^5$	$<10^5$	$<10^5$
Static decay						
Mil B-81705B	seconds	FTMS 101	<2.0	<2.0	<2.0	<2.0
NFPA Code 56	seconds	Method 4046	<0.5	<0.5	<0.5	<0.5
Flammability		Subject 94	V-O	V-O	V-O	V-O

Products According to Resin Types

Thermoplastic resins can be broken down into two general groups: amorphous and crystalline. General properties characteristic of either of the two groups can be applied to any thermoplastic compound.

The amorphous resins are characterized by toughness, good creep resistance, thermal stability, and the capability of being molded into complex shapes having tight tolerances. They have poor chemical resistance, do not reinforce to the same degree as crystalline resins, and are embrittled by the addition of fillers.

Compared with the base resin, crystalline resins reinforce well, with dramatic improvements in such physical properties as tensile strength, heat distortion temperature, and low mold shrinkage. Warp resistance, however, is poor for fiber-reinforced crystalline resins. Chemical resistance is superior, and fillers tend to embrittle crystalline resins. Crystalline resins can be made conductive more easily than amorphous resins. A lower concentration of conductive additive is required than that required for amorphous resins to achieve the same conductivity levels.

The thermoplastic resins generally used for conductive composites can be divided into the two categories as follows:

Crystalline Resins	Amorphous Resins
Nylon 6, 6–6, 6–10, 6–12, 11, 12	ABS
Polyesters—PBT, PET	Polystyrene
Acetal	SAN
Polyethylene	Polycarbonate
Polypropylene	Polysulfone
Polyphenylene sulfide	Modified PPO
PEEK	Polyethersulfone
	Polyetherimide
	Amorphous nylon
	Thermoplastic elastomers
	Polyurethane

With all the possible combinations of resins and conductive additives, it is no wonder that numerous commercial products are available. By first selecting the resin type that best suits the basic property requirements, then choosing a conductive product in this resin family with the desired conductivity, the product best suited for the intended application can be more easily determined. Tables 4.7 through 4.15 give examples of physical properties for crystalline and amorphous resins compounded with different types of conductive additives.

Nylon resins are characterized by their toughness, high mechanical

Table 4.7 Conductive Carbon Black–Filled Products Based on Crystalline Polymers

Property	Units	Test Method	Resin		
			Super-Tough 6–6 Nylon	PP	PE
Conductive category	—	—	Static dissipative	Static dissipative	Static dissipative
Specific gravity	—	—	1.16	1.09	1.04
Mold shrinkage	in./in.	D-955	0.016	0.017	0.022
Tensile strength	psi	D-638	8000	3000	3600
Tensile elongation	%	D-638	4–6	20	10+
Flexural strength	psi	D-790	12,000	3900	4300
Flexural modulus	psi × 10^6	D-790	0.4	0.15	0.150
IZOD impact	ft-lb/in. notch	D-256	1.5	3	1.2
Unnotched impact	ft-lb/in.	D-256	15	11	No break
HDT at 264 psi	°F (°C)	D-648	160	160	130
Surface resistivity	ohms/square	D-257	10^7	10^7	10^7
Volume resistivity	phms-cm	D-257	10^5	10^5	10^5
Static decay					
Mil B-81705B	seconds	FTMS 101	<0.05	<0.05	<0.05
NFPA Code 56	seconds	Method 4046	<0.5	<0.5	<0.5

Table 4.8 Conductive Carbon Black–Filled Products Based on Amorphous Polymers

			Resin			
Property	Units	Test Method	PC	Acetal	Polyester Elastomer 55D Hardness	Styrene Thermoplastic Elastomer 65A Hardness
Conductive category	—	—	Static dissipative	Static dissipative	Static dissipative	Static dissipative
Specific gravity	—	—	1.22	1.46	1.26	1.04
Mold shrinkage	in./in.	D-955	0.006	0.020	0.017	0.012
Tensile strength	psi	D-638	7000	8000	2600	800
Tensile elongation	%	D-638	2–4	6–10	10+	600+
Flexural strength	psi	D-790	11,000	12,500	1900	—
Flexural modulus	psi × 10^6	D-790	0.360	0.340	0.06	—
IZOD impact	ft-lb/in. notch	D-256	1.3	0.8	No break	No break
Unnotched impact	ft-lb/in.	D-256	12	8	No break	No break
HDT at 264 psi	°F (°C)	D-648	270	130	110	—
Surface resistivity	ohms/square	D-257	10^7	10^7	10^7	10^7
Volume resistivity	ohms-cm	D-257	10^5	10^5	10^5	10^5
Static decay						
Mil B-81705B	seconds	FTMS 101	<2.0	<2.0	<2.0	<2.0
NFPA Code 56	seconds	Method 4046	<0.5	<0.5	<0.5	<0.5

strength, high heat distortion temperatures, processing ease, and excellent low friction. Of the seven commercial types, nylons 6, 6-6, and 6-12 are the most common, with nylon 6-6 having by far the largest commercial usage. Both 6 and 6-6 nylon absorb moisture, which could limit their use in some applications. Where moisture absorption is a problem, low moisture absorption grades such as 6-10, 6-12, 11, and 12 nylons can be used. Of the nylons, 6 and 6-6 are the strongest structurally. A modified impact grade of 6-6 nylon, supertough nylon is the toughest, having notched and unnotched impact strengths of 15 and 30, respectively. On the other hand, the supertough grade of nylon has lower strength properties as well as lower thermal properties compared with 6-6 or 6 nylon.

The nylons, being crystalline, have the properties characteristic of crystalline polymers—good solvent resistance, high strength, and low mold shrinkage. Because of their anisotropic behavior, part warpage could be a problem, especially when combined with conductive fiber reinforcements.

A number of grades of conductive nylon products are marketed that use the various nylon properties. Although commercial carbon black–filled composites are available, products with the best strength properties contain PAN carbon fiber as the reinforcement. Properties for a number of conductive nylon products are listed in Tables 4.7, 4.9, 4.11, and 4.13.

Polypropylene and polyethylene resins have decreased physical strength when compared with the high-performance engineering thermoplastics, but they have low specific gravity, excellent chemical resistance, and good dimensional stability, surface finish, and ease of processing, and they are available in a wide range of melt flow grades at a cost below that of the other resin systems. The range of resin grades available for injection molding, extrusion, and blow molding applications permits conductive products to be made available for these applications.

Conductive carbon black is the most widely used of the conductive additives in commercial products.

Antistatic grades of conductive polymers predominantly use polypropylene and polyethylene resins because of the low processing temperature requirement. To enhance the mechanical properties, they are generally reinforced with fiberglass, particularly in combination with conductive carbon black. Fillers such as glass beads, mica, and talc are also combined with conductive carbon black in commercial products to improve stiffness and control warpage.

Of the classifications of polyethylene—high density, medium density, low density, and linear low density—high density is predominantly used for commercial products because of its higher physical properties and

Table 4.9 Pan Carbon Fiber–Reinforced Products Based on Crystalline Resins

			Resin				
Property	Units	Test Method	6–6 Nylon	PBT Polyester	PP	PPS	PEEK
% PAN carbon fiber	%	—	30	30	30	30	30
Specific gravity	—	—	1.33	1.42	1.04	1.45	1.42
Mold shrinkage	in./in.	D-955	0.001	0.001	0.001	0.001	0.001
Tensile strength	psi	D-638	38,000	27,000	8000	26,000	33,000
Tensile elongation	%	D-638	1–2	1–2	1–2	1–2	1–2
Flexural strength	psi	D-790	55,000	40,000	12,000	42,000	48,000
Flexural modulus	psi × 10^6	D-790	2.6	2.6	1.6	3.3	2.6
IZOD impact	ft-lb/in. notch	D-256	2.0	1.8	1.2	1.1	1.5
Unnotched impact	ft-lb/in.	D-256	20	15	2.5	6	12
HDT at 264 psi	°F (°C)	D-648	500	435	245	500	600
Surface resistivity	phms/sq	D-257	100	100	100	100	100
Volume resistivity	ohms-cm	D-257	5	1	5	5	<1
Static decay							
Mil B-81705B	seconds	FTMS 101	<0.05	<0.05	<0.05	<0.05	<0.05
NFPA Code 56	seconds	Method 4046	<0.5	<0.5	<0.5	<0.5	<0.5
Shielding effectiveness	decibels at 1000 MHz	ES7-83	30	—	—	—	—

Table 4.10 Pan Carbon Fiber–Reinforced Conductive Composites Based on Amorphous Resins

			Resin					
Property	Units	Test Method	PC	ABS	Acetal	Polysulfone	PEI	PES
% PAN carbon fiber	%	—	30	30	30	30	30	30
Specific gravity	—	—	1.32	1.18	1.46	1.35	1.38	1.48
Mold shrinkage	in./in.	D-955	0.001	0.002	0.001	0.001	0.001	0.001
Tensile strength	psi	D-638	26,000	18,000	14,000	25,000	32,000	28,000
Tensile elongation	%	D-638	1–2	1–2	<1	1–2	1–2	1–2
Flexural strength	psi	D-790	42,000	27,000	20,000	35,000	44,000	38,000
Flexural modulus	psi × 10^6	D-790	2.4	2.4	2.6	2.6	2.6	2.6
IZOD impact	ft-lb/in. notch	D-256	2.6	1.4	1.2	1.4	1.1	1.3
Unnotched impact	ft-lb/in.	D-256	15	7	3.4	8	9	10
DT at 264 psi	°F	D-648	300	230	—	365	410	415
Surface resistivity	ohms/sq	D-257	100	100	100	100	100	100
Volume resistivity	ohms-cm	D-257	1	1	1	1	1	1
Static decay								
Mil B-81705B	seconds	FTMS 101	<0.05	<0.05	<0.05	<0.05	<0.05	<0.05
NFPA Code 56	seconds	Method 4046	<0.50	<0.50	<0.50	<0.50	<0.50	<0.50

Table 4.11 Nickel-Coated Pan Carbon Fiber Products Based on Crystalline Polymers

			Resin					
Property	Units	Test Method	6–6 Nylon	6–6 Nylon	PPS	PPS	PEEK	PEEK
% PAN carbon fiber	%	—	20	40	20	40	20	40
Specific gravity	—	—	1.29	1.50	1.50	1.71	1.46	1.64
Mold shrinkage, 1/8″	in./in.	D-955	0.003	0.002	0.002	0.001	0.002	0.001
Tensile strength	psi	D-638	22,000	28,000	14,000	20,000	23,000	28,000
Tensile elongation	%	D-638	1–2	1–2	1	1	1–2	1
Flexural strength	psi	D-790	31,000	39,000	20,000	27,000	32,000	38,000
Flexural modulus	psi × 10^6	D-790	1.2	2.1	1.6	2.7	1.5	2.6
IZOD impact	ft-lb/in. notch	D-256	1.0	1.2	1.0	1.1	1.2	1.2
Unnotched impact	ft-lb/in.	D-256	19	10	3	4.5	10	14
HDT at 264 psi	°F	D-648	480	480	500	500	550	550
Surface resistivity	ohms/sq	D-257	100	100	100	100	100	100
Volume resistivity	ohms-cm	D-257	1	<1	1	<1	1	<1
Static decay								
Mil B-81705B	seconds	FTMS 101	<0.05	<0.05	<0.05	<0.05	<0.05	<0.05
NFPA Code 56	seconds	Method 4046	<0.20	<0.20	<0.20	<0.20	<0.20	<0.20
Shielding effectiveness	decibels at 1000 MHz	ES7-83	40–60	60+	40–60	60+	40–60	60+

Table 4.12 Nickel-Coated Pan Carbon Fiber Composites Based on Amorphous Thermoplastic Resins

Property	Units	Test Method	Resin		
			PC	ABS FR Grade	PES
% NCG fiber	—	—	20	20	20
Specific gravity	—	—	1.35	1.37	1.51
Mold shrinkage	in./in.	D-955	0.001	0.001	0.002
Tensile strength	psi	D-638	14,000	7000	17,000
Tensile elongation	%	D-638	1–2	1–2	1–2
Flexural strength	psi	D-790	18,000	11,000	28,000
Flexural modulus	psi × 10^6	D-790	1.1	1.2	1.5
IZOD impact	ft-lb/in. notch	D-256	1.1	1.2	1.0
Unnotched impact	ft-lb/in.	D-256	6	4.5	9.0
HDT at 264 psi	°F	D-648	290	220	415
Surface resistivity	ohms/sq	D-257	100	100	100
Volume resistivity	ohms-cm	D-257	<1	<1	<1
Static decay					
Mil B-81705B	seconds	FTMS 101	<0.05	<0.05	<0.05
NFPA Code 56	seconds	Method 4046	<0.50	<0.50	<0.50
Shielding effectiveness	decibels	—	40–60	40–60	40–60

Table 4.13 Aluminum Flake Filled Conductive Products Based on Crystalline Thermoplastic Resins

Property	Units	Test Method	Resin		
			6–6 Nylon	PP	PBT Polyester
Conductive category	—	—	Conductive	Conductive	Conductive
Specific gravity	—	—	1.48	1.23	1.66
Mold shrinkage	in./in	D-955	0.004	0.012	0.008
Tensile strength	psi	D-638	9500	3500	7500
Tensile elongation	%	D-638	2.0	3.0	1.5
Flexural strength	psi	D-790	16,500	5000	12,500
Flexural modulus	psi × 10^6	D-790	1.2	0.4	1.0
IZOD impact	ft-lb/in. notch	D-256	1.0	2.3	1.0
Unnotched impact	ft-lb/in.	D-256	4.5	6.0	4.0
HDT at 264 psi	°F	D-648	465	220	380
Surface resistivity	ohms/sq	D-257	100	100	100
Volume resistivity	ohms-cm	D-257	1	1	1
Static decay					
Mil B-81705B	seconds	FTMS 101	<2.0	<2.0	<2.0
NFPA Code 56	seconds	Method 4046	<0.5	<0.5	<0.5
Shielding effectiveness	decibels	ES7-83	40	35	40

Table 4.14 Aluminum Flake-Filled Conductive Products Based on Amorphous Resins

Property	Units	Test Method	Resin PC	Resin ABS FR Grade	Resin SMA
Conductive category	—	—	Conductive	Conductive	Conductive
Specific gravity	—	—	1.54	1.54	1.42
Mold shrinkage	in./in.	D-955	0.003	0.003	0.003
Tensile strength	psi	D-638	6400	3300	6500
Tensile elongation	%	D-638	2.0	1.5	1.8
Flexural strength	psi	D-790	12,500	6200	10,500
Flexural modulus	psi × 10^6	D-790	0.95	0.6	1.0
IZOD impact	ft-lb/in. notch	D-256	1.3	1.4	1.2
HDT at 264 psi	°F	D-648	290	190	266
Surface resistivity	ohms/square	D-257	100	100	100
Volume resistivity	ohms-cm	D-257	1	1	1
Static decay					
Mil B-81705B	seconds	FTMS 101	<2.0	<2.0	<2.0
NFPA Code 56	seconds	Method 4046	<0.5	<0.5	<0.5
Shield effectiveness					
at 1000 MHz, 1/8"	decibels	ES7-83	40	50	50

improved dimensional stability. Both copolymer and homopolymer polypropylene–based products are available, the properties of the conductive product depending upon the grade used. Tables 4.7, 4.9, 4.11, and 4.13 list products for several conductive grades of these products.

Of the thermoplastic polyester types of resin, polyethylene terepthalate (PET) and polybutylene terepthalate (PBT), PBT is the one available in commercial conductive composites. It is very highly crystalline and has properties similar to those of 6-6 nylon—toughness, high physical strength, high abrasion resistance, low coefficient of friction, and good surface appearance. PBT has low moisture absorption, and this property, along with the others listed previously, makes it a choice over nylon 6-6 for conductive applications. Physical properties for several conductive products made with PBT resin are listed in Tables 4.9 and 4.13.

Polyphenylene sulfide (PPS) and polyetheretherketone (PEEK) are crystalline resins with exceptionally high temperature properties, especially when reinforced. For conductive applications requiring high-strength and high-temperature properties, products containing either of these resins are good candidates. Because of the resin costs and to take advantage of the reinforced properties, products are available generally reinforced with carbon fiber for high-performance applications. Conductive carbon black grades can be made available but these tables list property data for selected compounds made from these resins.

Table 4.15 Stainless Steel–Filled Conductive Products Based on Amorphous Resins

			Resin			
Property	Units	Test Method	PC	PC FR Grade	PES	ABS FR Grade
Conductive category	—	—	Conductive	Conductive	Conductive	Conductive
Specific gravity	—	—	1.33	1.27	1.55	1.30
Mold shrinkage	in./in.	D-955	0.003	0.004	0.005	0.004
Tensile strength	psi	D-638	11,000	10,000	10,500	6000
Tensile elongation	%	D-638	4	5	1.7	4.0
Flexural strength	psi	D-790	17,000	16,000	18,900	11,000
Flexural modulus	psi × 10^6	D-790	0.5	0.45	0.6	0.43
IZOD impact	ft-lb/in. notch	D-256	1.4	1.3	1.2	0.6
Unnotched impact	ft-lb/in.	D-256	—	—	—	—
HDT at 264 psi	°F	D-648	285	285	420	210
Surface resistivity	ohms/sq	D-257				
Volume resistivity	ohms-cm	D-257	1	10	1	100
Static decay						
Mil B-81705B	seconds	FTMS 101	<0.05	<0.05	<0.05	<0.06
NFPA Code 56	seconds	Method 4046	<0.5	<0.5	<0.5	<0.3
Shielding effectiveness	decibels	ES7-83	50	40	50	40

Polycarbonate-based conductive products make use of this amorphous resin's key properties—toughness, dimensional stability, predictability, and isotropic mold shrinkage. The molding characteristics of the polycarbonate conductive products make these materials preferred over nylon 6-6 or other crystalline resins where solvent resistance is not a problem. Polycarbonate, being amorphous, has relatively poor solvent resistance. Conductive products are available containing a number of the conductive additives, alone and in combination with reinforcements and fillers to achieve improved property strengths. Properties for several grades are shown in Tables 4.8, 4.9, 4.12, 4.14, and 4.15.

The amorphous resins used in conductive products for high-temperature applications are polysulfone, polyethersulfone, and polyetherimide. These resins are classified as being high-temperature engineering thermoplastics and are characterized by having outstanding long-term creep resistance at elevated temperatures. Combining reinforcement with conductivity results in products with enhanced strengths that can be molded to close tolerances. Although conductive carbon black composites are available, these tend to have less toughness and are somewhat brittle compared with reinforced grades. They are generally used where there are high-temperature requirements but where low-strength properties are sufficient.

Polystyrene resins are hard and brittle, have good glass and high clarity, and can be easily processed. Antistatic formulations use the high-clarity and low-processing-temperatures requirement. Carbon black conductive additives tend to embrittle the resin and are unacceptable for most applications. Although carbon fiber–reinforced polystyrene has enhanced property strength as well as the necessary conductivity, poor solvent resistance and low heat resistance limit the applications other resin-based systems prefer.

ABS resins are characterized by their toughness, good surface appearance, and the ability to mold to tight tolerances. A wide range of impact grades is commercially available as well as several FR grades that find use in conductive products. Conductive carbon black tends to embrittle ABS, making these types of systems unacceptable because of their poor properties. Stainless steel fiber products have the properties of the base resin and are finding use in EMI/RFI applications. Tables 4.10, 4.11, 4.14, and 4.15 show properties for several ABS based resin products.

The properties of thermoplastic elastomers (TPE) allow for conductive products to be made available with good toughness and high flexibility, similar to rubberlike products. Mechanical properties vary broadly, depending on the type and grade of the resin. Hardness grades vary from 55 to 95 Shore A. Thermoplastic elastomers fall into four groups: thermoplastic polyolefins, thermoplastic polystyrene, thermoplastic polyesters, and thermoplastic polyurethanes. The thermoplastic polyesters have the best strength properties; the urethane grades, the best abrasion resistance; and the olefins, the best overall properties. Carbon black composites are generally the most frequently made available with TPE. Table 4.8 shows properties.

Melt-processable fluoropolymers offer a means to make conductive products available with extremely high chemical and thermal resistance properties. A number of grades of these fluoropolymers can be utilized:

ETFE, (ethylene tetrafluoroethylene)
FEP, (fluorinated ethylene propylene)
PFA, (perfluoralkory tetrafluoroethylene)
PVDF, (polyvinylidene fluoride)

Conductive products made with these resins must be processed at very high temperatures using corrosion-resistant metal for the process equipment.

Products According to Applications

ESD-sensitive devices must be protected as they are being transported or stored. Protection begins at the initial stage of production and continues

Table 4.16 Melt-Processable Fluoropolymers Containing Carbon Fiber

Property	ASTM Test Method	Resin	
		ETFE	PFA
Specific gravity	D-792	1.74	2.08
Mold shrinkage			
in./in., 1/8 in. section	D-955	0.018	0.030
Impact strength, IZOD ft. lb./in.			
notched, 1/8 in.	D-256	3.5	No break
unnotched, 1/8 in.	—	No break	No break
Tensile strength, psi	D-638	5500	3200
Tensile elongation, %	D-638	27	20
Flexural strength, psi	D-790	5400	4100
Flexural modulus	D-790	180,000	150,000
Volume resistivity, ohm-cm	D-257	10^4	10^2
Surface resistivity, ohms/sq	D-257	10^4	10^2
Deflection temp., °F,			
at 264 psi	D-648	240	270
Static decay			
Mil B-81705B, 2.0 sec	FTMS-101B Method 4046	0.30	0.30
NFPA code 56A, 0.5 sec	FTMS-101B Method 4046	0.06	0.06

through the various subassembly stages to the shipping and storage of the final assembly. Packages would include such items as bags, tote boxes, DIP tubes, chip carriers, and storage boxes and bins and must provide protection against static charge buildup from electrostatic fields and against direct discharge from contact with a charged object. Because of the many different requirements—strength properties, thermal properties, solvent resistance, conductivity range and cost, to name a few—one type of material cannot meet all requirements. It becomes necessary to choose the most cost-effective systems from the wide range of conductive materials to protect ESD-sensitive devices properly under all the environmental conditions to which they may be exposed.

ESD protection application requirements are generally divided into three categories—conductive (less than 10^5 ohms/square surface resistivity), static dissipative (10^5–10^9 ohms/square surface resistivity), and antistatic (greater than 10^9 but less than 10^{14} ohms/square surface resistivity). These categories provide a guideline for choosing a material for an application.

In the following section, various applications are described covering the types of products that are suitable candidates for the application.

A. ESD Protective Bags, Sheet and Film, DIP Tubes or IC Rails (Table 4.17)

Bag products provide shielding protection for storing, handling, or transporting expensive circuit boards, microchips, and explosive chemical mixtures. The products are a protective measure against electrostatic discharge, chemical cling, and explosive chemicals. Transparent and colored bags with either static decay or static dissipative properties are available in polyethylene-based resins. Black, opaque bags with surface resistivity of less than 10^5 ohms/square are produced from conductive carbon black polyethylene composites. Conductive film grades are available that fall into either of the three conductive classifications and are used to form bags or simply to wrap material for storage, handling, and shipping.

Chair mats, tabletop covers, floor and wall covering, and thermoformed parts such as trays, bins, tote boxes, tops, and funnels are several applications for conductive sheets. Transparent high- and low-density polyethylene have antistatic and static dissipative properties. Polypropylene as well as high- and low-density polyethylene conductive carbon black–filled composites are extruded into black, opaque sheet with surface resistivity less than 10^5 ohms/square.

DIP tubes or IC rails provide a method of protecting dual in-line microprocessors from both electrostatic or mechanical damage in shipping, storage, and in-plant handling. They can be used with automatic

Table 4.17 ESD Protective Bags, Sheet and Film, Dip Tubes and IC Rails

Resin	Classification	Additive	Comments
PP/PE	Antistatic	Organic antistat	1. Transparent (translucent) and colorable 2. Low cost 3. Nonpermanent, may require periodic testing and cleaning 4. Effectiveness depends on moisture or relative humidity 5. Many grades available with a wide range of mechanical properties and melt flow 6. Will not slough, acceptable for clean room applications 7. Cleaning solutions or buildup of dirt, oil, etc., will affect performance 8. Will not conduct 9. Will not accept printing 10. Only antistatic range of conductivity available
PE/PP	Static dissipative	Ionic additive	1. Transparent, can be colored 2. Permanently antistatic 3. Nonsloughing, acceptable for clean room applications 4. Will not conduct 5. No effect on base polymer properties
PP/PE	Antistatic static dissipative conductive	Carbon black	1. Opaque black 2. Higher cost 3. Permanent 4. Effectiveness independent of moisture or relative humidity 5. Has sloughing characteristics, not acceptable for clean room applications 6. Can conduct, must be used with care near electrical connectors or other electrical sources 7. Will accept printing 8. Many grades available with a wide range of mechanical properties and melt flow 9. Three classifications of conductivity available

(continued)

Table 4.17 ESD Protective Bags, Sheet and Film, Dip Tubes and IC Rails *(continued)*

Resin	Classification	Additive	Comments
PS	Antistatic	Organic antistat	1. Very transparent, can be colored 2. Improved stiffness compared to antistatic grades of polypropylene and polyethylene 3. Impact grades available but product is opaque 4. Nonpermanent 5. Has similar properties described for items 3–10 for antistatic polypropylene and polyethylene

insertion machines, and being transparent, they facilitate the identification of the IC circuits enclosed. Clear or colored tubes are available in the antistat and static dissipative classification. Opaque, black tubes, containing conductive carbon black, fall into the conductive classification. DIP tubes or IC rails are generally extruded using either polypropylene, polyethylene, or polystyrene. Injection molding grades as well as extrusion grades are available.

B. ESD Protective Tote Boxes, Lids, Vertical Dividers, Compartment Trays, and Storage Bins (Table 4.18)

These applications provide a means of storing and transporting sensitive devices, protecting them from ESD damage. Tote boxes, compartment trays, and storage bins are manufactured in various sizes, in a number of colors, and in the various classifications of conductivity. As with the DIP tubes and bags, only antistatic and static-dissipative transparent or colored products are available.

Opaque products with surface resistivity values less than 10^5 ohms/square contain conductive carbon black. These highly conductive tote boxes, when properly covered with a conductive lid, can provide a Faraday cage for parts. Table 4.18 lists some of the benefits and limitations for the products generally used for these types of applications. Compounders can tailor-make products to meet specific requirements for a tote box application by combining reinforcements and/or fillers with the conductive additive. In addition, grades can be selected from the large variety of commercial polypropylene homopolymer and copolymer resins to achieve a desired property such as a specific melt flow or a particular izod impact value.

Polypropylene is the resin of choice for most applications. Where

Table 4.18 ESD Protective Tote Boxes, Lids, Vertical Dividers, Compartment Trays, and Storage Bins

Resin	Classification	Conductive Additive	Comments, Benefits, and Limitations
PP/PE	Antistatic	Organic antistat	1. Economically priced 2. Transparent with colors available 3. Solvents can temporarily affect efficiency 4. Controlled blend-off of charge when adequately grounded 5. Humidity dependent
PP/PE	Antistatic Static dissipative	Conductive carbon black	1. Provides a Faraday cage for parts when container is covered 2. Fast static decay 3. Independent of humidity or solvent 4. Sloughing characteristic—not acceptable for clean room applications 5. Black only color
PP/PE	Static dissipative	Conductive carbon black with reinforcement	1. Improved physical properties 2. Reduced mold shrinkage 3. Tailor-made products with specific properties available 4. Other properties similar to item 2
PP/PE	Static dissipative conductive	PAN carbon fiber	1. Greatly improved physical properties 2. Low mold shrinkage 3. Very high cost

high physical strength properties are needed, PAN carbon fiber is the conductive additive.

C. ESD Protective IC Chip Carriers (Table 4.19)

IC chip carrier applications vary widely in their requirements—cost, high physical strength, flame retardancy, solvent resistance, thermal properties, processing ease, and dimensional stability, to name a few. No one product can meet all the requirements, and chip carriers are injection molded from a variety of resin families, each combined with one or more of the conductive additives. Table 4.19 shows a selection of conductive products for this application.

Carbon black polypropylene carriers are the most economical, have good chemical resistance, and can be processed easily. Tailor-made prod-

Table 4.19 ESD Protective IC Chip Carriers

Resin	Classifications	Additives	Comments
PP	Static dissipative conductive	Conductive carbon black or milled carbon fiber or pitch carbon fiber	1. Low cost 2. Reinforced or filled grades available with improved mechanical strength properties 3. Easily processed 4. Good chemical resistance 5. FR products available
PP	Static dissipative conductive	PAN carbon fiber	1. High cost 2. Very high mechanical properties compared to conductive carbon black composites 3. Good chemical resistance
PC	Static dissipative conductive	Carbon black	1. Medium cost 2. Molds to dimension, good warp resistance 3. Could be attacked by some solvents 4. FR composites available 5. Reinforced grades available
PC	Static dissipative conductive	PAN carbon fiber	1. High cost 2. High-strength properties 3. Poor chemical resistance 4. Good thermal strength properties
Nylon 6–6	Static dissipative conductive	Conductive carbon black Pitch carbon fiber Miller carbon fiber	1. Medium cost 2. Good processing characteristics 3. Reinforced and impact modified grades available 4. FR grades available 5. Good solvent resistance
Nylon 6–6	Static dissipative conductive	PAN carbon fiber	1. High cost 2. Very high mechanical strength properties 3. FR grades available 4. Good thermal properties
PES PEI Poly-sulfone	Static dissipative conductive	PAN carbon fiber	1. Very high thermal properties 2. Good mechanical strength properties 3. Fair solvent resistance 4. High cost 5. Molds to close tolerances

(continued)

Table 4.19 ESD Protective IC Chip Carriers *(continued)*

Resin	Classifications	Additives	Comments
PPS	Static dissipative conductive	PAN carbon fiber	1. Very high thermal properties 2. Good mechanical strength properties 3. Excellent chemical resistance 4. Anisotropic shrinkage behavior 5. High cost
PPS	Static dissipative conductive	Conductive carbon black	1. Lower cost compared with PAN carbon fiber composites 2. Reinforced grades available 3. High thermal properties 4. Anisotropic shrinkage behavior 5. Average solvent resistance
FEP, PFA	Antistatic static dissipative conductive	Conductive carbon black	1. Very high cost 2. Very high chemical resistance 3. Very high thermal properties 4. Requires special corrosion resistant process equipment
FEP, PFA	Static dissipative conductive	PAN carbon fiber	1. Very high cost 2. Very high chemical resistance 3. Very high thermal properties 4. High mechanical strength properties 5. Requires special corrosion resistant process equipment

ucts are available that use reinforcements or fillers to alter properties. They have relatively low physical and thermal properties.

Chip carriers based on polycarbonate resin use this resin's amorphous characteristics—toughness, isotropic mold shrinkage, and low mold shrinkage. Where additional physical strength is required, reinforcements can be combined with the conductive additive. Polycarbonate, on the other hand, has poor chemical resistance.

The crystalline nature of nylon and PBT imparts high strength, rigidity, heat resistance, and chemical resistance to the chip carriers. Greatly enhanced properties are achieved with reinforcements.

For chip carriers requiring very high thermal properties, conductive products compounded from polyethersulfone (PES), polyetherimide (PEI), polysulfone, or polyphenylene sulfide (PPS) are available.

PES, PEI, and polysulfone are amorphous; PPS is crystalline. The properties of chip carriers based on these resins reflect either their amorphous or crystalline character. Besides these properties, the products have high-thermal and high-strength properties, but they are also expensive.

For the ultimate in thermal properties and chemical resistance, melt-processable fluoropolymers such as Teflon PFA and Teflon FEP resins can be used for chip carriers. Conductive products of these two resins are extremely tough, have excellent thermal properties and excellent solvent resistance and are inherently flame retardant. Very high process temperatures are required as well as special corrosion-resistant screws, barrels, and molds. Fluorocarbon conductive products are very expensive, but these resins perform under environmental conditions that rule out other polymers.

D. EMI–RFI Shielding Composites for Housings and Enclosures (Table 4.20)

Aluminum flake–filled thermoplastic housings are the most cost effective but they have low mechanical strength and poor surface finish. For some applications, painting may be necessary, for aesthetic reasons. A major benefit with aluminum flake is its very high thermal conductivity.

PAN carbon fiber and nickel-coated PAN carbon fiber greatly enhance the strength properties of housings and enclosures, at the same time providing shielding effectiveness values greater than 40 dB. Colors are available with the nickel-coated graphite systems; only black is available with the PAN carbon fiber.

Stainless steel fiber composites used for EMI/RFI applications retain the properties of the base resin, can be colored, provide a good surface finish, and are relatively easy to mold. Proper mold design and control over molding, however, is required to achieve the optimum shielding effectiveness for molded parts.

Test Methods

By far the most widely accepted test method to determine the conductivity of plastics and plastic compounds is ASTM D-257. This test method defines the procedure to use to measure surface resistivity and volume resistivity. For thermoplastics intended to be used to dissipate electrostatic charges, surface resistivity is the more meaningful of the two, ESD being a surface phenomenon.

Despite the prevalent use of ASTM D-257, there are problems associated with it. The test method was initially developed for testing insulative materials, not static-dissipative or conductive materials. It was also designed to test a large diversity of materials, resulting in procedures for

Table 4.20 EMI/RFI Applications

Resin	Conductive Additive	Comments
ABS	Aluminum flake	1. Shielding effectiveness values greater than 40 dB 2. FR composites available 3. Mottled finish, may need painting for aesthetics 4. Lowest cost 5. Low mechanical properties but excellent thermal conductivity
ABS/PC Alloy PC Nylon 6–6, 6, ST-801 ABS	PAN carbon fiber	1. Shielding effectiveness values greater than 40 dB 2. FR composites available 3. High cost 4. Very high mechanical strength properties 5. Black color only
ABS PC Nylon 6–6, 6, ST-801	Nickel-coated graphite fiber	1. Shielding effectiveness values greater than 40 dB 2. FR composites available 3. Very high cost 4. High mechanical strength properties 5. Colors available
ABS PC Nylon 6–6, 6, ST-801	Stainless steel	1. Shielding effectiveness values greater than 40 dB 2. Requires low loadings, little to no effect on properties 3. Good surface finish 4. Can be colored 5. FR composites available

11 different electrode systems. These problems are now being addressed. Perhaps a more appropriate test method will be issued in the near future. ASTM D-257 still remains the preferred test method for measuring the electrical characteristics of a plastic material, and published data for conductive plastics are based on this test method.

Surface Resistivity

Surface resistivity as measured by ASTM D-257 is numerically equal to the surface resistance between two electrodes forming opposite sides of a square. Surface resistivity values are reported in ohms/square.

The size of the square is immaterial, but it is extremely important that the electrode make good contact with the sample surface. Simply touching the surface with point electrodes of an ohmmeter is unacceptable. Infinite resistivity readings may result when actually the sample is conductive. A test set up that applies a constant pressure such as 500 psi also insures that good contact will be made.

A number of different electrodes can be used to measure surface resistance, as described in D-257. One such device is the guard ring electrode shown in Figure 4.1. The measurement for it is performed by applying a voltage from a DC power supply to the unguarded electrode and connecting an electrometer to the guarded electrode. The electrometer must have an impedance of at least 10^{14} ohms. The surface resistivity is calculated using the equation from ASTM D-257. See Figure 4.2 for description of dimensions.

$$e_s = \frac{R_s P}{g}$$

$$R_s = \frac{V}{I} = \text{surface resistance (ohms)}$$

V = test voltage in volts

I = current in amps

$P = \pi D_0$ = perimeter (cm)

$$D_0 = \frac{D_1 + D_2}{2} = \text{average diameter (cm)}$$

$$g = \frac{D_2 - D_2}{2} = \text{sample width (cm)}$$

Another type of electrode configuration is one in which the electrodes are 1.00 ± 1 in. in length and are separated 1.00 ± 1 in. by a nonconductor. A test area of 1 in.^{2} results when placed on the surface of the

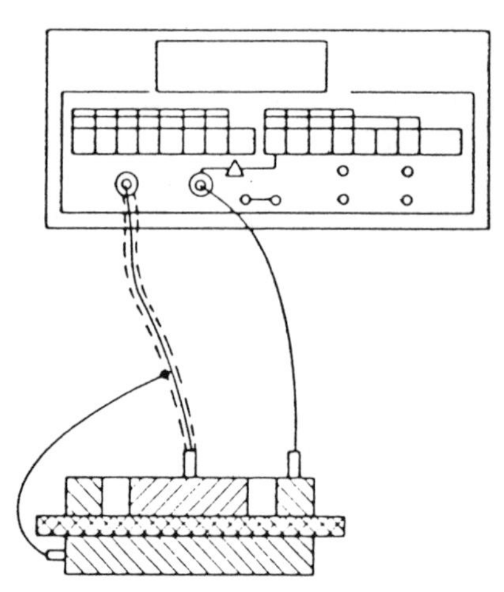

Figure 4.1 Surface-resistance measurement with guard ring electrode.

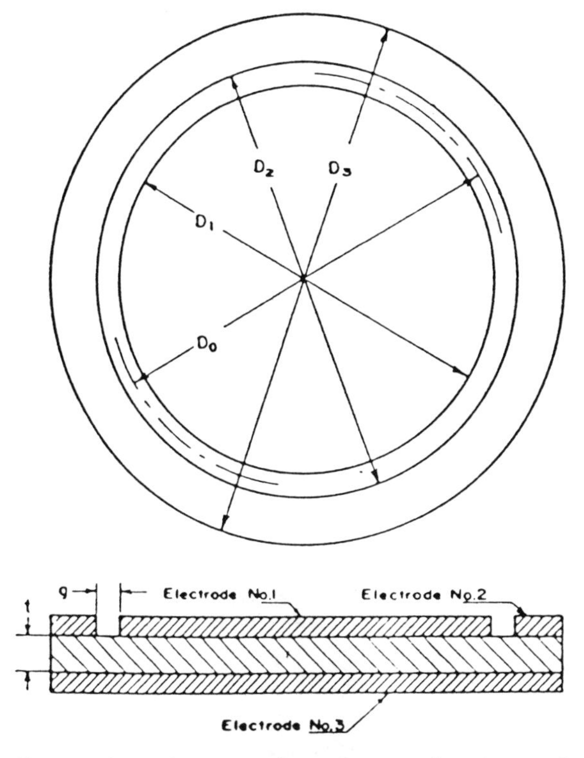

Figure 4.2 Flat specimen for measuring volume and surface resistance.

specimen. An ohmmeter is used to measure the surface resistivity, which is the same as the resistance between the electrodes. The electrodes may be designed to give a point contact by making them at an angle. This test jig is normally weighted to ensure good contact with the specimen. Figure 4.4 shows this electrode configuration.

For surface resistivity readings below 10^6 ohms/square, an ohmmeter such as a Simpson 260, amprobe instruments AM-2A, or other similar types with the same output voltage of 1.5V DC will give accurate readings in this conductivity range.

Above 10^6 ohms/square, it is preferred to use a 500-V DC output voltage meter such as a Hiptronics megohmmeter HM3A. The higher voltage gives more accurate readings by eliminating most contact resistance problems associated with higher resistance measurements.

There are a number of commercially available, direct-reading surface resistivity meters that offer simplicity and convenience for both routine and investigative resistance measurements. These instruments are simple to use, are portable, and provide reliable measurements in the 10^4–10^{14} ohms/square range. Two examples of portable, direct-reading surface

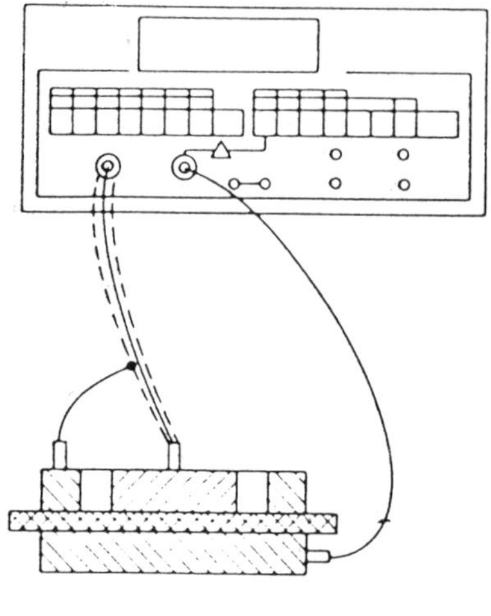

Figure 4.3 Volume resistance measurement with guard ring electrode.

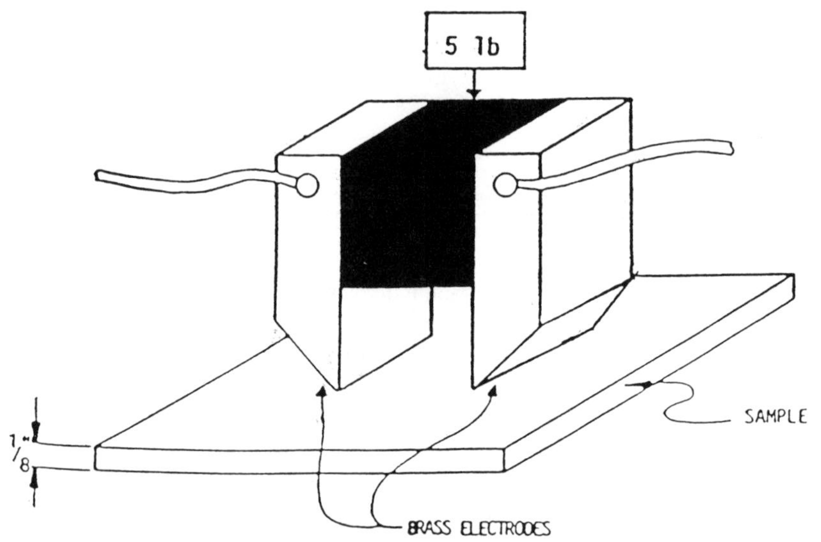

Figure 4.4

resistivity meters are the SRM-110 supplied by Voyager Technologies, Inc., and the Model 262 supplied by Monroe Electronics, Inc.

The Voyager SRM 110 weighs 2 lb. and has a test voltage of 30 V and a contact pressure of 40 lb/in.^{2}/probe. Surface resistivity measurements can be made up to 10^{12} ohms/square. The highest possible conductivity reading is 10^5 ohms/square. Any higher conductivity will only be indicated as less than 10^4 ohms/square. The Voyager SRM-110 functions by applying 30 V to each of three outer probes with a ground return provided by a center, spring-loaded probe. The outer probes are 1.5 in. from the center probe. Each of the probes is made from a conductive elastomer to ensure proper contact with the surface under test. To use, simply place the instrument on the surface to be measured, turn on, and read the LED indicator. The value for surface resistivity is rounded off to the nearest integer. No additional pressure is required to ensure good contact; the weight of the instrument in conjunction with the conductive elastomer probes is sufficient.

The Monroe Model 262 has a test voltage of 100 V for readings above 10^5 ohms/square and automatically switches to 10 V for the 10^4 and 10^5 ohms/square readings to protect the instrument. It has a range of 10^4–10^{14} ohms/square. Two metal rails that have compressed strips of conductive elastomeric material embedded in them are located on the base of the instrument. In use when the power is turned on, voltage is applied to one of the rails, and the current flowing between the rails is detected by the other one. The internal circuit converts the signal into a direct readout of ohms/square. The contact rail, which is 5 in. long, requires a test specimen to be at least this long for accurate measurements. Test specimens less than 5 in. long can be tested but with a loss in accuracy.

Because of the number of possible electrode configurations and test methods that can be used to measure surface resistivity, it is extremely important that the material supplier and customer communicate about the test method each uses. If the material supplier uses a method different from his customer's incoming QC test method, material rejections may take place.

Volume Resistivity

Volume resistivity (also referred to as bulk resistivity) is, as described in D-257, numerically equal to the volume resistance in ohms between opposite faces of a 1-cm cube of material. As with surface resistivity, a number of electrode systems can be used for its determination.

The guard ring electrode described earlier for the determination of surface resistivity can also be used to determine volume resistivity. The electrode configuration is shown in Figure 4.3. Using Figure 4.2 for the description of dimensions, volume resistivity is calculated as follows:

$$e_v = A/t\, R_v$$

e_v = volume resistance (ohms-cm)

$A = \frac{\pi (D_1 t_g)^2}{4}$ = area of the electrodes (cm^2)

R_v = volume resistance between electrodes

$g = \frac{D_2 - D_1}{2}$ = sample thickness (cm)

Guard ring electrodes are commercially available so that both surface and volume resistivity measurements can be made simply by changing the connection to the electrodes.

Another test method is to paint a 1.3-cm band of conductive silver paint completely around each end of a test bar 15 × 1.3 × 0.3 cm to form electrodes (as per ASTM D-257, Section 6). The volume resistance between the electrodes is determined by attaching the ohmmeter leads to the silver electrodes and the resistance rendered. Volume resistivity is calculated using the equation

$$e_v = A/L\, R_v$$

where R_v = measured volume resistance (ohms)
A = area of the effective electrode
L = distance between electrodes

Static Decay Test

The static decay test measures the ability of a material, when grounded, to dissipate a known charge that has been induced on the surface of the material, the Test Method Standard 101B—Method 4046, "Antistatic Properties of Materials."

The static decay procedure is used by the government in Mil. B-81705B, which specifies acceptable decay times for packing materials for electrosensitive devices and explosives. The military standard states that a material must dissipate an induced charge of 5000 V completely within 2 sec at a relative humidity of 15% or less.

The National Fire Protection Association (NFPA) also adopted the FTMS 101B, Method 4046 static decay test in its Code 56A, a standard that defines the requirements for materials to be used in hospital environments. Code 56A calls for the 5000-V charge to drop to 10% of its initial value at 50% relative humidity within 0.5 sec.

An Electro-Tech Systems' Model 406C static decay meter is a completely integrated system for measuring the electrostatic properties of materials in accordance with Method 4046 and is considered the standard test equipment in the industry today for measuring static decay.

A humidity test chamber is required in conjunction with the static decay meter to control the humidity for performing the test under the specified humidity conditions. The humidity chamber is an airtight acrylic glove box. Moist air within the chamber is dried with a circulating pump and desiccator. Access to the chamber is through a square door opening on one side held in place by four latches. Samples placed in the chamber are handled by the operator placing his hands in the neoprene gloves located in the front. Dusting the hands lightly with talc aids in inserting and removing the hands from the gloves.

In operation, two desiccators are used for drying, one inside the test chamber, the other outside the test chamber connected in series with a pump and the chamber. The use of the outside desiccator speeds up the process of obtaining low humidity within the chamber by circulating the air within the chamber through the desiccator and back into the chamber. Once the desired relative humidity is reached, the internal desiccation will maintain the level without having to use the pump continuously.

The test procedure to measure static decay with the Electro-Tech Systems' Model 406C Static Decay Meter and humidity chamber follows:

1. Test specimens are molded into 5 × 3 × 1/8 in. plaques.
2. The test plaques are conditioned for 24 hours at 73°F and 15% relative humidity for NFPA 56A.
3. Using the neoprene gloves, insert the test specimen into the holder located in the Faraday cage of the static decay meter, close the cage, and test for any initial charge.
4. Apply a 5000-V positive charge and record the charge accepted after one minute.
5. Discharge the applied voltage and note the time required to dissipate to 0 (zero) voltage at 15% relative humidity if testing against Mil. B-81705B or to 500 V at 50% relative humidity if testing against NFPA 56A requirements.
6. Repeat the test for a total of three times for a positive 5000-V charge and three times for a negative 5000-V charge.
7. Average the six dissipation times for the final value.

EMI/RFI Shielding Effectiveness Test Method

ASTM ES7-83, entitled "Electromagnetic Shielding Effectiveness of Planar Materials," sets forth two different methods for testing EMI shielding materials. It is important to note that ES7-83 is only intended to rank materials, and the results are not necessarily absolute. Open field sight measurement is the only way to determine the actual shielding effectiveness of a complete product.

Systems for shielding applications are tested for attenuation or reduction of a signal that is expressed in decibels (dB). The units (decibels) are not linear but logarithmic. The formulation for attenuation of shielding effectiveness (SE) is:

$$SE = 20 \log E_1/E_2$$

where E_1 = the preshielded signal intensity in V/M
E_2 = the postshielded signal intensity in V/M

Table 4.21 shows levels of shielding effectiveness versus signal intensity reduction. The higher the attenuation, the better the shielding effectiveness. The minimum shielding effectiveness for most applications is 30 dB.

Electromagnetic radiation has two components consisting of an electric field and a magnetic field, both of which can cause interference in circuits within a specified range. The characteristics of an electromagnetic wave changes with distance from the source and frequency (wavelength) of the source. If the distance from the source is greater than the wavelength/ 2π, the region is described as far field. If the distance is less than the wavelength/2π, the region is described as near field.

ASTM ES7-83 describes a method to use for each of these two regions—a transmission line method for far field data and a dual-chamber method for near field data.

Shielding effectiveness also will vary over the frequency range for any given system. The test method as described in ASTM ES7-83 specifies testing at four frequencies: 30 MHz, 100 MHz, 300 MHz, and 1000 MHz.

The test equipment consists of four components: a signal generator, two 6-dB attenuators, a receiver, and a sample fixture. The importance of using properly shielded cables and connectors cannot be overemphasized. A block diagram of the test system is shown in Figure 4.5. Detailed drawings for the shielded box and the transmission line sample holder are included in the ASTM specification. These can be purchased from several commercial sources.

The dynamic range of the test equipment must first be measured. This is done by using a test specimen such as an aluminum panel that is opaque to radio frequency. This would be the maximum shielding effectiveness that can be measured by the system. The standard states that test results cannot be reported any higher than 3 dB less than the dynamic range. The dynamic range of the equipment is to be reported along with test results. The factors that can limit the dynamic range are the shielding

Table 4.21 Attenuation vs. Signal Intensity Reduction

Attenuation (dB)	Signal Reduction
10	90.9
20	99.0
30	99.9
40	99.99
50	99.999
60	99.9999
70	99.99999

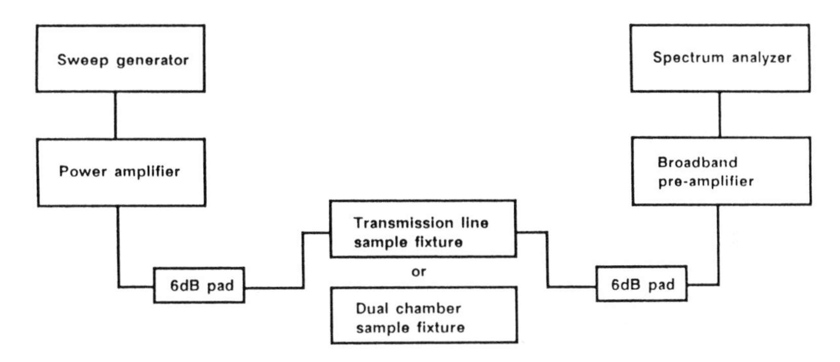

Figure 4.5 Block diagram of test system.

of the coaxial cable and connectors and the maximum output power of the generator.

In addition, a standard should be measured for shielding effectiveness and the value obtained should be within ±3 dB of the known value. The standard is a gold-metallized plastic sheet of proper surface resistivity.

For the dual-chamber method, the specimen size required is 76.2 mm wide, 152.4 mm long, and 3.2 mm thick (3 × 6 × 1/8 in.). To ensure good contact with the sample holder, the standard requires that a conductive paint be applied to the perimeter of the specimen.

For the transmission line method, the test specimen is a "washer" with an outside diameter of 99.75 mm (3.93 in.), an inside diameter of 43.70 mm (1.72 in.), and a thickness of 3 mm (0.125 in.). For best electrical contact, the specimen should be machined to size to remove the resin-rich surface and conductive paint applied to the abraded surface. The specimen should be tested as soon as possible after the paint dries.

Although not a direct measurement of shielding effectiveness, volume resistivity can provide a rough estimate of the expected shielding effectiveness for a material. At least 1 ohm/square is required to be acceptable as a candidate. A plot of shielding effectiveness versus volume resistivity for a typical system is shown in Figure 4.6. It is important to note that for some systems, although high readings for volume resistivity are obtained, the material may still be an effective shield for EMI/RFI.

Injection Molding of Conductive Materials

Processing methods and mold design of conductive materials, as with any thermoplastic compounds, are determined by the resin type, conductive additive type, and additive level.

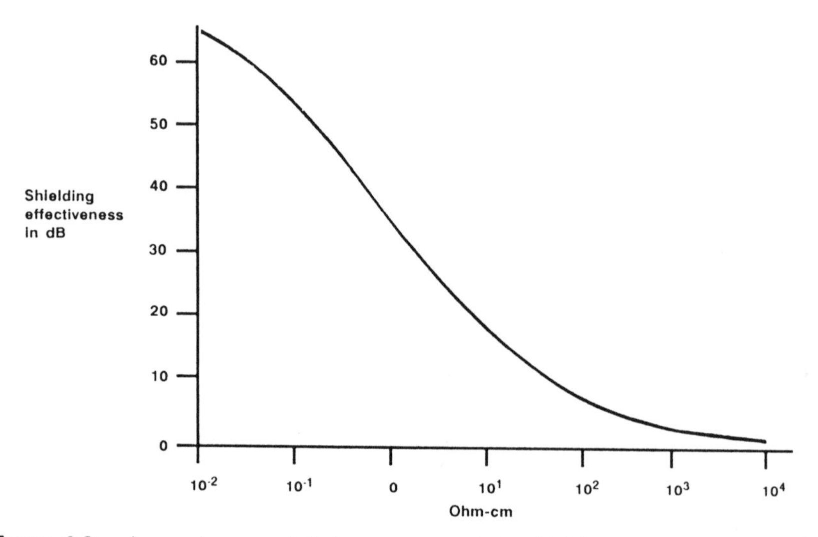

Figure 4.6 Log volume resistivity versus average shielding, 0.1 to 1,000 MHZ.

The resin type is the predominant factor in setting the general processing parameters.

The additive type can have considerable effect on processing variables and recommended mold design. Conductive additives can be fiber reinforcement types or particulate filler types.

A fiber reinforcement additive is one that has a high aspect ratio between the length and the diameter. Conductive materials can be made with reinforcement additives such as carbon fibers, nickel-plated carbon fibers, metallized glass fibers, or stainless steel fibers. To maintain the compound's physical characteristics and conductivity level during processing, the critical aspect ratio of the fiber must be preserved. This critical length-to-diameter ratio is generally considered to be 20:1. The specific type of fiber will determine its ability to flex rather than break during molding and, consequently, will determine how sensitive the processing must be.

A particulate filler additive generally has a very low aspect ratio. Fillers do not improve most physical properties. Typical fillers are carbon blacks and metallic flakes and particles. They have wide processing parameters since aspect ratio is not a concern.

The third factor affecting processing is the additive level. Conductive compounds may contain anywhere from 3% to 60% additive, depending on the type of additive and the conductivity required.

For discussion purposes, comparisons will be made between conductive compounds and unfilled and/or glass-filled resins. Since most data have been developed on carbon fiber and carbon black compounds, the following information pertains primarily to these conductive additives.

Gate and Runner Design

Usually by the time a molder receives a mold, the configuration of the cavity is set. How and where to gate the parts, however, including runner size and configuration, is frequently the molder's decision.

The path that a resin takes between the injection molding machine barrel and the part cavity can have a real effect on finished part physicals and conductivity.

The runners required for conductive compounds need to be larger than those for unfilled resins and similar to those for glass-filled materials. This is especially true for fiber-filled conductives. The number of bends or turns should be minimized to prevent fiber breakage, as previously discussed.

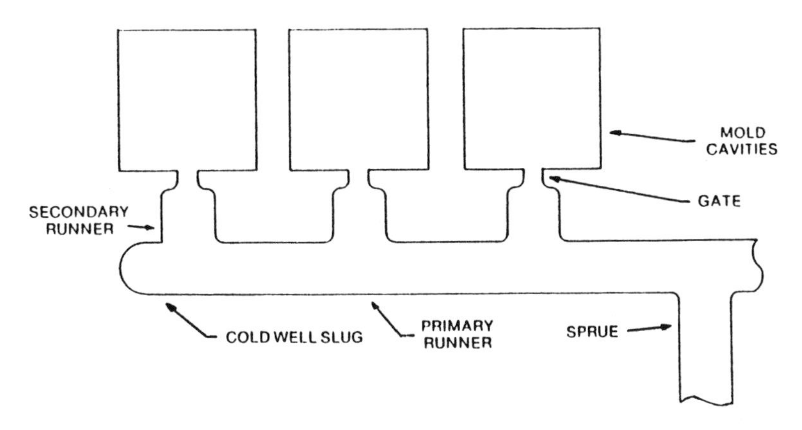

Figure 4.7

Cold well slugs are recommended at the end of 90 angles. Full round runners are recommended for optimum flow.

The gating type, size, and location constitutes one of the most important factors in obtaining quality parts. This is also the item most frequently overlooked or misapplied. All conductive compounds will require gates larger than unfilled resins because of higher viscosities and fiber damming. Fiber damming, like a river log jam, occurs when too many fibers try to move through too small an opening. This results in severe pressure drops and unfilled or unpacked parts.

Guidelines for gating of conductive-filled materials are to have a minimum gate depth of 0.060 in., with a preference for a 0.090- to 0.100-

in.-deep gate. Fiber-filled compounds require gating of the larger size. Because of this gate size requirement, a pinpoint, tunnel, or subgate are not recommended. A tab or sprue gate is preferred.

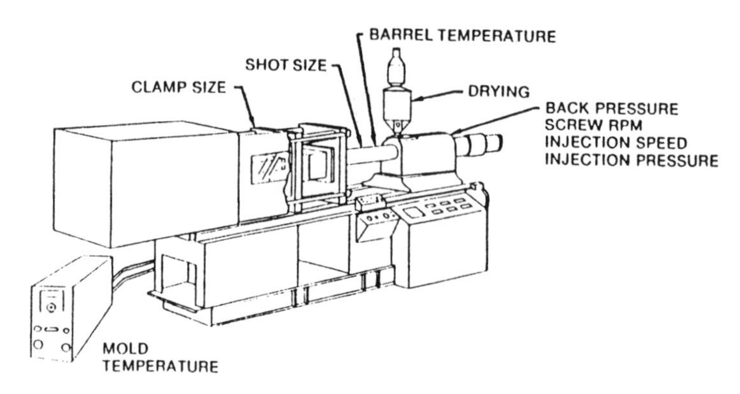

Figure 4.8

Machine Sizing

As with any molding job, the correct size of the injection molding machine must be chosen. Three criteria must be considered—mold size, projected area resulting in clamp tonnage required, and shot size.

The molding machine must have adequate platen area and spacing between tie bars to allow the mold to be mounted and clamped in place.

The projected area of the cavities and runners must be calculated to determine the clamp tonnage required to mold parts without flash. The clamp tonnage recommended per square inch of projected area for carbon black–conductive materials should be 0.5–1 ton/in.^{2} higher than the recommended clamp tonnage of the same resin unfilled. A carbon fiber–filled material should be calculated at 1–2 tons/in.^{2} higher than its unfilled resin.

The shot size requirement would be calculated as with any resin, with the recommended shot being 40% to 70% of barrel capacity.

Processing Parameters

The processing variables associated with the injection molding operation can be broken down into seven areas. These are not all independent functions, since changing one parameter may have an effect on another. For simplicity, however, we will discuss suggested settings and the reasons for these. The seven parameters are

Drying	Injection pressure
Screw speed	Barrel temperature

Injection speed
Back pressure
Mold temperature

Drying

Straight unfilled resins, depending on their type and whether they are hygroscopic or not, may or may not require drying. Once fillers or reinforcements are added to thermoplastic resins, drying becomes important. The fillers and especially fiber reinforcements can cause a "wicking" action whereby moisture is absorbed into the resin pellet by capillary action along the filler or fiber. The normally hydroscopic resins of course must be dried. The presence of moisture can cause a severe loss of physical properties. The temperature and time required depend on the base resin.

Recommended Drying Time

		Drying	
RTP Series	Polymer Type	Temp (°F)	Time (hr)
100	Polypropylene	175	2
200	Nylon 6/6	*175	4
200A	Nylon 6	*175	4
200B	Nylon 6/10	175	4
200C	Nylon 11	175	4
200D	Nylon 6/12	*175	4
200E	Amorphous nylon	175	4
200H	Nylon impact modified	175	4
300	Polycarbonate	*250	4
400	Polystyrene	*180	2
500	SAN	*180	2
600	ABS	*180	2
700	Polyethylene	175	2
800	Acetal	200	2
900	Polysulfone	250	4
1000	PBT	*250	4
1100	PET	*275	4
1200	Polyurethane thermoplastic elastomer	*225	6
1300	Polyphenylene sulfide	250	2
1400	Polyethersulfone	*300	6
1500	Polyester thermoplastic elastomer	*200	4
1600	Polyphenylsulfone	*250	4
1700	PPO (modified)	200	2
2100	Polyetherimide	*300	4
2200	Polyetheretherketone	*300	3
2700	Styrenic thermoplastic elastomer	175	2
2800	Olefinic thermoplastic elastomer	175	2
3400	Liquid crystal polymer	300	8

Screw Speed

The screw RPMs will affect shear on the resin and consequently the heat or melt temperature. Our biggest concern, however, is the effect of the mixing and tumbling action, which can cause fiber breakage below the critical aspect ratio. This will cause lower physical properties and conductivity. Carbon fibers are more brittle than glass fibers; this causes them to break more readily when they undergo a bending force. It is therefore suggested that the screw RPMs for a carbon fiber–conductive resin be set at the minimum speed that will not delay the molding cycle. The carbon black conductives do not have this problem and will tolerate the higher screw RPMs.

Injection Speed

Most filled resins require a fast fill or injection speed to obtain an appearance surface. Again, the fiber-filled resin with a fast fill will be subjected to forces that tend to break the fiber as they go through the runners and gates. For this reason, trying to run the slowest speed without sacrificing appearance is the recommended approach.

The surface conductivity also depends on how resin-rich the surface skin is. A faster fill increases the resin richness, producing lower surface electrical conductivity.

Back Pressure

Back pressure, as with screw speed and injection speed, affects the shear rate the material is subjected to. Because the material supplier should do a good job of compounding, there is no need for the additional mixing that high back pressure provides. Back pressure should only remove air from the melt stream. The suggested setting for back pressure is to just obtain a reading on the gauge, such as 25–50 lb. A higher setting for filler-type additive materials is acceptable but should not be required.

Injection Pressure

The injection pressure is very dependent on part geometry, runner and gate configuration, melt temperature, mold temperature, and viscosity of resin. The pressure should be set to provide a fully packed cavity without flash. In general, because the conductive additives increase the resin viscosity, a higher injection pressure will be required with these than unfilled resins.

Barrel and Mold Temperature

The barrel temperature and mold temperature are variables where real differences in the quality of the molded conductive parts can occur. With conductive materials, two inherent properties affect these settings. One is the melt viscosity of the compound. The melt viscosity increases with increasing conductive additive. A higher viscosity means it flows and pushes harder through the screw and into the runner and mold cavity.

As an example, compare the melt indices of various filled polypropylene materials. The lower melt index numbers translate into higher viscosities, which means they are harder to flow. These compounds were produced using a nominal melt index polypropylene of 6 g /10 min. The melt index test was run at 230 C and 2160 gram weight.

Compound system	Melt index (Grams/10 min)
Unfilled polypropylene	6.0
Low carbon black–filled PP	3.0
High carbon black–filled PP	0.5
10% glass-filled PP	2.0
40% glass-filled PP	0.5
10% carbon fiber–filled PP	1.5
40% carbon fiber–filled PP	>0.5

Carbon black–filled materials exhibit viscosity changes similar to those of glass-filled products. Carbon fiber–filled materials have a higher viscosity than similarly filled glass products. The difference in carbon fiber versus glass fiber is likely due to the actual volume loading percentage being higher with carbon than that with glass fiber, since these percentages are on a weight basis. Carbon fiber specific gravity is lower than glass fiber, so at a given weight percent the volume percent is greater.

These melt indices are for a given temperature. A corrective action to obtain better flow is to raise the barrel temperatures and raise the mold temperatures.

The second inherent property of conductives is thermal conductivity. Generally, the electrical conductive medium also increases the thermal conductivity of these compounds in proportion to the percentage of fill. As a comparison, consider the thermal conductivity of various filled polypropylenes. The conductivity is measured in BTU/hr/FR^2/ F/in.

Compound system	Thermal Conductivity
Unfilled polypropylene	0.8
Low carbon–filled PP	1.4
High carbon–filled PP	1.8
10% glass-filled PP	1.3
40% glass-filled PP	2.5
20% carbon fiber–filled PP	3.8
40% carbon fiber–filled PP	4.7

As the thermal conductivity increases, the ability of the material to absorb or pick up heat improves, as does the ability to give off or lose heat. This effect means that the resin will reach a melt point quicker in the injection molding barrel and that the material will solidify faster when injected into a cold mold.

To counteract this change may require lowering barrel temperatures and increasing mold temperatures.

The compound viscosity is a combination of the resin viscosity at a given melt temperature and the filler level. To lower the base resin carrier viscosity, it is necessary to compensate with higher barrel temperatures.

The following graphs show the effect of barrel temperatures on viscosity and thermal conductivity.

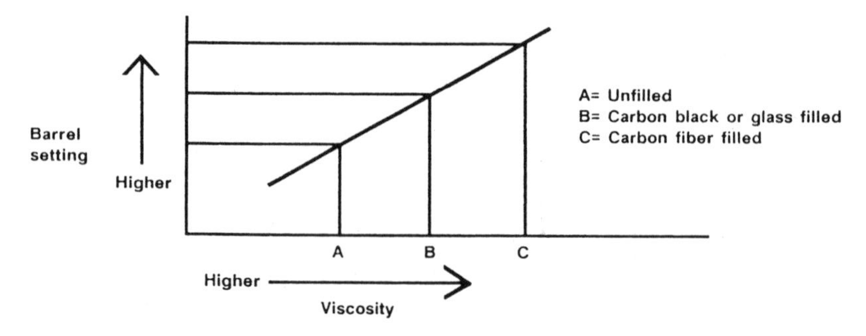

Figure 4.9 Barrel temperature vs. viscosity.

Since the compound viscosity is a combination of the resin viscosity at a given melt temperature and the filler level, the need to compensate with higher barrel temperatures is required to lower the base resin carrier viscosity.

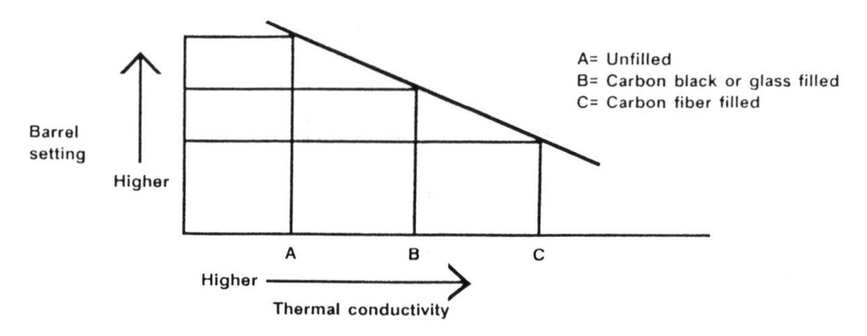

Figure 4.10 Barrel temperature vs. thermal conductivity.

Because the increase in thermal conductivity causes the compound to reach the required melt temperature quickly, the barrel temperature does not have to be set as high to drive the heat into the material.

Since these two inherent material properties require barrel compensation in opposing directions, the resultant barrel setting change is to some degree canceled.

A guideline to the required compensation, taking this into effect, is as follows:

1. A carbon black or glass-filled product will require slightly higher barrel settings than an unfilled resin of the same type.
2. A carbon fiber–filled material will require a lower temperature setting than either a glass-filled or unfilled resin, as the thermal conductivity more than compensates for the viscosity increase.

Following is a graphic representation of the effect on mold temperature based on viscosity and thermal conductivity:

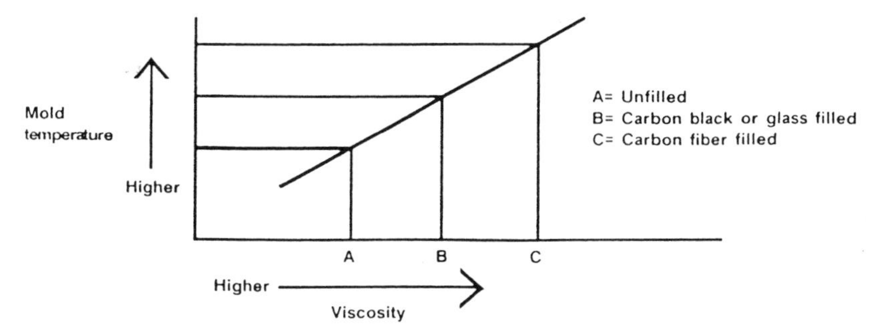

Figure 4.11 Mold temperature vs. viscosity.

As the viscosity increases, it becomes harder to push resin through the runner and mold cavity. Raising the mold temperature keeps the resin melt higher for a longer period of time.

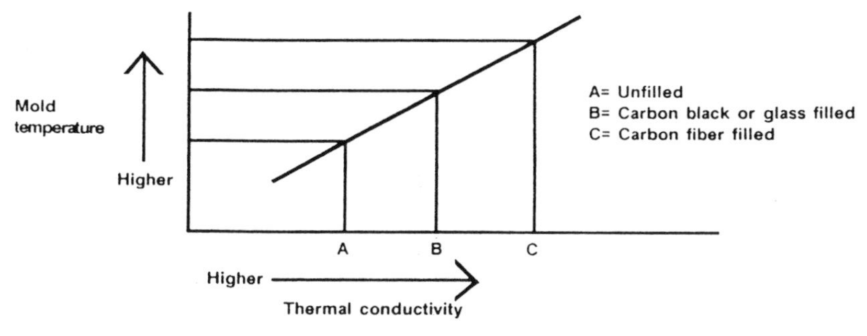

Figure 4.12 Mold temperature vs. thermal conductivity.

As the thermal conductivity increases, the melt front tends to solidify faster because the cold mold causes poor fill and short shots. Increasing the mold temperature will provide slower cool and prevent premature solidification.

Since these two inherent material properties require mold temperature compensation in the same direction, their effect is additive.

A guideline to the required total compensation required in establishing the mold temperature is as follows:

1. A carbon black–filled or a glass-filled material will require higher mold temperatures than a nonfilled resin of the same type.
2. A carbon fiber–filled resin will require higher mold temperatures than either of the preceding.

Typical Mold and Barrel Temperature for Conductive Compounds

	Mold (°F)		Barrel (°F)	
Resin	Carbon Black	Carbon Fiber	Carbon Black	Carbon Fiber
Polypropylene	140	—	460	—
Nylon 6/6	220	240	520	500
Nylon 6	180	200	470	450
Polycarbonate	200	240	550	530
ABS	—	200	—	440
Polyethylene	100	—	400	—
Polysulfone	250	300	670	650
PBT	—	200	—	470
PPS	—	280	—	590
PEI	—	300	—	690
PEEK	—	350	—	740
Styrenic thermoplastic elastomer	100	—	350	—

Regrind Consideration and Usage Regrind usage is always a consideration for economic reasons. General guidelines for unfilled resins usually indicate acceptability of 20% to 30% regrind back with virgin material.

Conductive compounds that are based on filler additives, such as carbon black materials, are acceptable for regrind in the same percent usage—20% to 30%. Our only concern is for the degradation of the base resin.

A carbon fiber–filled product, however, is susceptible to both resin degradation and fiber breakage. The resultant loss in physical properties is greater than that of unfilled resins at the same regrind level. The additional loss of properties because of fiber breakage indicates that many carbon fiber materials cannot tolerate regrind. It is our recommendation that the product and process be tested for the level of regrind that will be acceptable.

Summary

The processing of conductive materials will depend on the additive used. Conductive additives can be reinforcement (fiber) or filler (particle) types. The particle-filled (carbon black) materials are much less sensitive and the primary variable for unfilled resins is the need for a hotter mold.

The fiber-filled conductive materials require essentially the same processing parameters as glass-filled compounds of the same base resin, with the preventative alteration to minimize fiber breakage being even more important. In addition, because of the higher thermal conductivity, a hotter mold is required than would be necessary with a glass-filled resin.

Summary of Guidelines for Processing Conductive Materials

1. Gating:

	Filler Additive	Fiber Additive
Minimum gate depth:	0.060″	0.090″
Preferred gate depth:	0.090″	0.100″
Gate pinpoint	Not recommended	Not recommended
Tunnel	Not recommended	Not recommended
Subgates	Not recommended	Not recommended
Tab or sprue gate	Preferred	Preferred

2. Machine sizing:

Mold size—adequate platen area and spacing between tie bars to allow the mold to be mounted and clamped in place.

Shot size—recommend 40% to 70% of barrel capacity

Clamp tonnage:

	Filler Additive	Fiber Additive
Clamp tonnage, per in.^{2} of projected area	1/2–1 ton/in.^{2} higher than the unfilled grade	1–2 ton/in.^{2} higher than the unfilled grade

3. Drying: Recommended for conductive compounds

4. Processing speeds, pressures, temperatures:

	Filler Additive	Fiber Additive
Screw speed	Higher RPMs	Minimum speed w/o causing delays in molding cycle

(continued)

Summary of Guidelines for Processing Conductive Materials *(continued)*

4. Processing speeds, pressures, temperatures: *(continued)*

	Filler Additive	Fiber Additive
Injection speed	Can be faster	Slowest speed w/o sacrificing appearance
Back pressure	Can be higher	Low—25–50 lb
Injection pressure	Must be higher because of higher viscosities	
Barrel temp	Slightly higher than unfilled	Slightly lower unfilled
Mold temp	Higher than unfilled	Higher than any other grade
5. Regrind usage	20% to 30%	Not recommended

Part III

Index

Index

---

* PPO designates two different chemicals, in separate chapters, as indicated in the text.